PRAISE FOR CARLOS MAGDALENA'S

The Plant Messiah

"It is fascinating to follow Magdalena as he travels from remote Australian billabongs full of rare water lilies to the dry forests of western Peru. . . . But the greater accomplishment of *The Plant Messiah* is the compelling case that Magdalena makes [for] caring about plants in general."
—*Los Angeles Times*

"An entertaining, planterly romp."   —*Financial Times*

"A gripping account. . . . The work done by Magdalena and others like him is nothing short of miraculous. . . . [*The Plant Messiah*] illustrates just how much can be done to save even species that all but the greatest optimist would consider doomed."   —*The New York Review of Books*

"Bound to enthrall. . . . With evident passion for endangered and common plant species alike, Magdalena . . . shares his experiences traveling the world in his quest to save plant species before they go extinct."   —*Publishers Weekly*
(starred review)

"Magdalena's claim to be a green-fingered Jesus has its merits. His record of propagating plants that others have consigned to botanical history is impressive. . . . Some 30,000 plants have recorded uses for humans. Most people, the messiah preaches, are blind to these everyday miracles. This book will teach them to see."   —*The Economist*

CARLOS MAGDALENA

The Plant Messiah

Carlos Magdalena is the Tropical Senior Botanical Horticulturist at the Royal Botanic Gardens, Kew, and an international lecturer. He is renowned for his unique skills as a plant propagator who is saving the world's rarest plants.

THE

Plant Messiah

ADVENTURES IN SEARCH OF THE
WORLD'S RAREST SPECIES

CARLOS MAGDALENA

ANCHOR BOOKS
A Division of Penguin Random House LLC
New York

*To my mother, Edilia, who instilled in me a deep love
of nature, and to my son, Matheo. May I pass on to
him that same passion, so that one day he may enjoy
opportunities similar to those that have enriched my life.*

FIRST ANCHOR BOOKS EDITION, MARCH 2019

Copyright © 2017 by Carlos Magdalena

All rights reserved. Published in the United States by Anchor Books, a division
of Penguin Random House LLC, New York. Originally published in hardcover
in Great Britain by Viking, a division of Penguin Random House Ltd., London,
in 2017, and subsequently in hardcover in the United States by Doubleday,
a division of Penguin Random House LLC, New York, in 2018.

Anchor Books and colophon are registered trademarks of
Penguin Random House LLC.

The Library of Congress has cataloged the Doubleday edition as follows:
Name: Magdalena, Carlos, author.
Title: The plant messiah : adventures in search of the world's
rarest species / by Carlos Magdalena.
Description: [New York] : Viking, 2017. |
Includes bibliographical references.
Identifiers: LCCN 2017027303
Subjects: LCSH: Plant conservation. | Personal narratives.
Classification: LCC QK86.A1 M34 2018 |
DDC 333.95/316—dc23
LC record available at https://lccn.loc.gov/2017027303

**Anchor Books Trade Paperback ISBN: 978-0-525-43666-9
eBook ISBN: 978-0-385-54362-0**

*All interior illustrations are by Lucy Smith
Book design by Pei Loi Koay*

www.anchorbooks.com

Printed in the United States of America
10 9 8 7 6 5 4 3 2 1

Look again at that dot. That's here. That's home. That's us . . .

The Earth is the only world known so far to harbor life . . . Like it or not, for the moment the Earth is where we make our stand . . .

There is perhaps no better demonstration of the folly of human conceits than this distant image of our tiny world. To me, it underscores our responsibility to deal more kindly with one another, and to preserve and cherish the pale blue dot, the only home we've ever known.

—CARL SAGAN, reflecting on a photograph of Earth taken by the *Voyager 1* space probe from a distance of 6 billion kilometers (*Pale Blue Dot*, 1994)

Contents

❊❊

Prologue

I stood in front of the greenhouse bench. It was a frosty early morning at the Royal Botanic Gardens, Kew, London.

In front of me was a café marron plant, a gorgeous shrub that never stops blooming, with dark green leaves and snow-white jasmine-like flowers. It had been grown from cuttings taken from a plant on the island of Rodrigues in the Indian Ocean.

Actually, I should say *the* plant: it was the last of its species left in the world. This species, with the Latin name *Ramosmania rodriguesi*, had long been thought extinct. It hadn't been seen in the wild for more than fifty years, but suddenly, in 1980, it was rediscovered by a schoolboy.

Cuttings have limited uses. Only seed production could secure its long-term survival in the wild. Without seeds, it was destined to die. Experts had tried everything over the years to make this happen, but had failed.

Now it was my turn. Could I break the code?

I chose a bloom, then carefully unwrapped the scalpel blade. I placed it against the flower and held my breath.

I was about to make the cut that could change the fate of this species.

Introduction

Let me introduce myself. My name is Carlos Magdalena, and I am passionate about plants.

In 2010 I was labeled "El Mesías de las Plantas" by Pablo Tuñón, a journalist who wrote about my work in the Spanish newspaper *La Nueva España*. I suspect the name was partly inspired by my post-biblical (though pre-hipster) beard and long hair, and also because I was spending a lot of my time trying to save plants on the brink of extinction.

This alias reached a worldwide audience when Sir David Attenborough mentioned it while interviewing me for *Kingdom of Plants*, a series filmed at the Royal Botanic Gardens, Kew. The Plant Messiah quickly became my media moniker, offering ample opportunity for mockery among friends and colleagues alike. My family love the idea of my mum coming onto the balcony to shout, "He's not the Messiah, he's a very naughty boy!" in the style of the legendary sketch in Monty Python's *Life of Brian*.

Don't panic, though. I don't have a messiah complex.

I recently looked up the word *messiah*. It has several definitions: "a leader regarded as the savior of a particular country, group or cause," "a zealous leader of some causes or project," "a deliverer," and "a messenger." I'm aiming to be them all.

My mission is to make you aware of exactly how important plants are; in fact, I am obsessed with this idea. I want to tell you all about them and what they do for us, how crucial they are for our survival, and why we should save them. Plants are the key to the future of the planet—for us and our children—yet they are taken for granted by billions every day and we are often dismissive of their benefits. I am frustrated, sometimes angry, at this ignorance and indifference.

We may be blind to the fact, but plants are the basis of *everything*, either directly or indirectly. Plants provide the air we breathe; plants clothe us, heal us, and protect us; plants provide our shelter, our daily food, and our drink. Think medicines, building materials, paper, rubber for car tires and contraceptives, cotton for denim jeans and linen for dresses. Think bread, beans, tea, orange juice, beer, and wine. Think Coca-Cola. Then think about how cows eat grass, silage, or hay, supplying our meat and milk; how chickens eat wheat and seeds and give us eggs; how sheep eat grass and give us wool.

You see, plants are our greatest yet most humble servants; they care for us every day, in every way. Without them we would not survive. It is as simple as that.

In return for their generosity, we treat them appallingly. They are unappreciated and drastically undervalued. We don't treat them like servants, but like slaves. Their homes are destroyed, their families decimated. They are forced to mass-produce and are sprayed with chemicals. Factory farming is not just for animals but for plants also, and the environmental cost of this can be just as destructive (unsustainable palm oil development is just one sad example of this).

We destroy rainforests to plant crops in soil that can't support them. Without thought of what treasures the forests might hold, we drive flora and fauna to critically endangered levels and even extinction. During exploration and colonial expansion, we introduced goats to islands where they duly grazed the unique and delicate native flora until there was none left, removing the "green glue" that stabilized the soil and causing erosion problems that washed entire islands away. We introduced invasive weeds: a stifling, creeping death, smothering the local flora in a sinister form of botanical colonialism. Even today we build houses on agricultural land, laying endless miles of lifeless, white-lined tarmac over what were once wild, flowery meadows, blocking our minds to the consequences. It's a display of "plant blindness" of epidemic proportions. With the destruction of plants, we destroy the fauna too. Bird, mammal, and insect species—all gone forever. We rarely even think about what we're doing, and when we occasionally do, we still don't fully understand the consequences.

We have moved away from millennia of direct contact with plants; since the Industrial Revolution the majority of the population in developed nations have never worked with them and rarely communed with them. In the shift from the countryside to the city, we have lost our direct link with plants.

A great part of the problem is that whatever we do to them, plants can't speak, they can't plead their cause, warn of the folly of their destruction, or remind us of their importance with a raised voice or slammed fist on a table. Plants do not bleed when slashed, can't scream when burned. They can't write a message in a book. They need someone to do it for them.

If they can't produce seeds to ensure their survival, because their populations are so badly fragmented or reduced, or the survivors are just clinging to life, they need someone who will speak up on their behalf. They need someone who will say, "I will not tolerate extinction." Someone who understands plant science and will passionately champion their cause, using whatever means possible to ensure their survival.

Many of the world's great gardens, like Kew, are not just

for public education and entertainment. They collect and conserve rare species, in cultivation and in the wild, saving them from oblivion and making them available to science, and have done so for generations. The collective academic and horticultural genius of their staff is unrivaled, their collections world-renowned. Though they are committed and passionate, they need someone to broadcast their message to the people on a global scale.

I want to be this person.

I want to make the world aware of what plants do for us. I want us to give plants a value and appreciate what they do. I want us to understand their importance for our survival and the survival of our families—our babies, grandparents, and future generations. I want us to realize that without them we would die, and most living things on land and in the air would die with us. I want us to be enthused by the importance of conservation, to be fired with the determination that we should never give up, even if there is only one plant left in the world. I want us to understand the importance of plants so much that we are moved to do something about it.

A messiah can't change attitudes without supporters who will spread the gospel. When it comes to conservation we need passion, we need motivation, and we need action. It is time for a change.

I want this book to start that change. People need plants and plants need people, and spreading that message begins with me and you.

1

Genesis

To understand what motivates a plant messiah, you need to understand my heritage.

I was born in 1972, in the little town of Gijón in Asturias, northern Spain. I must have inherited my taste for working the land, and my love of flowers, from my mother, Edilia, a florist.

Though my sister and my brothers are interested in the natural world too, I am the only one of us who makes a living from it. My sister, Claudia, the oldest of my siblings, works in a Spanish version of Harrods. My older brother Falo, who was a salesman, sadly died five years ago. Another brother, Miguel, is a truck driver, and another, Javi, runs a small music club. I am the youngest. Like any large family, we have a range of talents—there's the sporty one, the artistic one, the musician, the naturalist. I have always been able to learn something from them, and from my uncles, aunties, and cousins too. I have definitely been shaped by the interests, passions, and fears of our tribe.

My mother was nine at the start of the Spanish Civil War, and her family suffered greatly. There were often insurrections in Asturias. In 1934 there was a miners' strike, when the anarcho-syndicalists declared themselves independent from Spain. It developed into a revolutionary uprising that was crushed by General Franco, who deployed Moorish troops to do so. Today the uprising is often considered to be the prelude to the Spanish Civil War, which itself is often considered to be the spark that ignited the Second World War, so it could be argued that the Second World War started in Asturias too.

The civil war had a massive impact on people's lives, as bitter lines were drawn between Republicans and Fascists. It was even fought within families: you could end up shooting your uncle, or your father, without realizing it. My mother lived in torment from the age of nine to about thirteen, and in the post-war period there was heavy rationing and a right-wing Catholic dictatorship to endure. Then war exploded again, this time throughout the rest of Europe. These were not ideal circumstances for any young girl or boy to grow up in.

After the civil war, she and her seven brothers and sisters worked the land, but the food was taken by the state and they had little left for themselves. My grandfather used to grow a small amount of tobacco, and he hid it in the maize fields so that the army wouldn't confiscate it, but they always somehow found out.

It was hard for the family: they had very little food and barely any resources. Everyone had to be self-sufficient, but not in the modern, fashionable sense of the word—this was for real. It was the only way they could survive.

Franco and his supporters had a fascinating attitude toward nature. They wanted to homogenize the country and eradicate anything that threatened productivity. In previous centuries, people had felled large expanses of ancient oak forest in Asturias and other regions in northern Spain, some of the most biodiverse places in Europe. Much of this wood was used to construct the galleons that first reached the Americas and then went on to form the "invincible armada" of the Spanish navy. Franco continued logging these rich native forests, and made the problem worse by replacing native species with rows and rows of eucalyptus and pines. It was like an ethnic cleansing of nature.

As a consequence, even today, Spain catches fire every summer. The state blames it on people having barbecues or throwing cigarettes from their car. Is it their fault, though? Or is it the fault of Franco and company, who destroyed the richly diverse flora and fauna and planted highly flammable vegetation instead? There is now a movement to clear out the eucalyptus and replace them with native species, but it is hugely expensive and you have to remove every single eucalyptus stump because the trees regrow vigorously when coppiced.

Many villages, including San Esteban de Dóriga, where my mother lived, were surrounded by forests that had been there since the Iron Age. You could keep bees there, harvest berries and mushrooms, graze your cows and goats. These native forests were a useful resource to the whole community, year after year. You couldn't "slash and burn" the whole lot, but you could cut down a tree and bring it into the village for your own use.

Franco, however, wanted to populate the whole of Spain, and put everything to practical use. Any animal that did not produce a profit was vermin and had to be destroyed. People would go out shooting in the forest, put the "unproductive" dead bears and wolves in a car trunk, then go to the town center and show them for a government fee. Records for the whole of Spain show that in 1969, 150 bears were killed. By the 1980s, when I was a kid, there were only eighty left.

These records make for sobering reading.[1] From 1944 to 1961, the total number of birds, mammals, and reptiles killed in Spain was tallied at 655,010. Among them were 1,206 golden eagles, 11,105 black kites, 47,739 ravens, 2,278 choughs, 103,322 magpies, 1,961 wolves, and 10,896 snakes.

Poisoning was by far the most destructive method. Vultures suffered because people would put out meat laced with strychnine as bait for other animals, and their corpses, which the vultures fed on, would become toxic too. They forgot that vultures stopped the spread of disease (if a cow dies of contagious bovine tuberculosis, for example, vultures will pick the bones clean, preventing the disease from spreading to other animals). Perhaps people thought that God made the land and the vermin so we could kill them for entertainment.

Although Franco's policies drastically reduced the numbers of wild animals, thankfully none resulted in mass extinction.

Yet we have not learned from our mistakes. Even today, farmers demand that the authorities keep killing wolves, even though when this is done in a random way it is detrimental to farming. Breaking up wolf packs causes even more damage to farmers, since lone wolves are more likely to attack livestock, which are easy prey. And most of the damage attributed to wolves is actually caused by wild dogs, which, in turn, are one of the wolves' favorite prey. Funny, that.

Hearing these stories when I was young made me aware of the importance of ecosystems and how vital it is to conserve animals and plants. I became interested in politics and started to realize that the wanton destruction of nature was part of the folly of man.

❖❖

Sandwiched between the Cantabrian Mountains and the sea, Asturias is one of the most rewarding places on earth—if you are into natural history. It is about thirty miles wide at one end, and about twelve at the other, and the topography is steep. Rivers tumble straight down from the mountains into the sea. You can be at 1,500 miles above sea level, looking at the rugged mountain scenery, yet only about twenty miles from the sea. Among the peaks there are waterfalls and several glacial lakes. It is one of the few places that openly display the geological history of the earth, from the day when the first molten rock solidified. In one place there are dinosaur footprints; in another, mounds of coral fossils, or fern fossils in carbon deposits.

Asturias is an unbelievable place for wildlife—the perfect place to learn about nature as a child. There are about seventy protected areas (landscapes, natural reserves, and national natural monuments), and the first national park ever declared in Spain, the Picos de Europa National Park. The jagged limestone mountains of the Picos de Europa define the eastern part of the region. They are intricate, with steep, narrow valleys and gorges, sometimes running north to south, then suddenly east to west. It is like a corrugated fingerprint of valleys, so a valley that is two and a half miles away in a straight line might be six miles away by road. What's more, Asturias boasts the largest patch of primary deciduous forest in Europe, the last viable population of brown bears and the largest wolf population in Western Europe, not to mention the greatest densities of otters, boars, and chamois on the continent.

Near to where I spent my childhood are the River Nalón and its main tributary, the River Narcea. The Nalón drainage system flows from the mountains through pristine forest and is teeming with salmon and aquatic wildlife (sometimes I think that I am like a salmon, born in the northern Spanish rivers and migrating to England). I call the Nalón the Amazon of Asturias. Just as the Amazon has its Río Negro (Black River), so too did the Nalón: when I was a child, the waters in the mid-lower course of the river were like dark chocolate, it was so heavily polluted with charcoal washings from the charcoal production there, but a program to restore the water quality has helped enormously.

The village where my mother was raised, San Esteban de Dóriga, near the River Narcea, had only about thirty inhabitants when she was growing up. It is surrounded by forests, hedgerows, and apple orchards, and although it is part of Spain, it is far from sun-bleached: Asturias has almost double the annual rainfall of London.

Asturias was the kind of place where people would work together for the benefit of the community. If there was a road to be built or the forest needed clearing to stop forest fires, everyone joined forces and gave their time for nothing. I am sure life in Britain was once like this too. The traditions and landscape of Asturias, with its corners of almost untouched land, deeply affected my attitude toward wild habitats and their conservation.

About twenty miles from where I lived was an industrial town called Avilés. It was, and still is, highly polluted and was recently claimed to be the most polluted town in Spain. When I was young, you could smell the town five miles before you reached it, and if you were unlucky enough to have to go there, you always ended up with streaming eyes or a hacking cough. In 1980, a report in the newspaper *El País* stated that six out of ten people seen in the local hospital emergency room were there because of respiratory problems such as chronic bronchitis.

It seemed incredible that you could have both of these scenarios within twenty miles of each other. On one hand, there was rich biodiversity, wildlife, and rugged scenery, and on the other an industrial, polluting nightmare, suffocating the life around it. All that was good and bad on the earth was there. Having seen both up close, I knew which side I wanted to choose.

❧ ❧

By the time I was five, I was looking after the plants at school and had become the authority on natural history for my friends. If I didn't know the answers to their questions I would go home and ask my mother or look in books until I found out. When I was older, I read all six volumes of an encyclopedia called

Natural Science, from cover to cover—twelve times. As they were expensive, my father said I should read them only if I rested them on a table, but I would take them to the bathroom, lock myself in, and sit there for hours. My family still has them.

Natural history soon became my sole passion and interest. I knew the names of all the fish in our aquarium, the birds that flew around the town, and the plants in the fields and surrounding streets. Back then we didn't have the same focus on television and social media as children do now; to keep ourselves entertained we played with our dogs or looked after parrots or exotic birds or went walking in the surrounding countryside. I learned how to care for my pets, looked up the places where they came from, and studied how they lived. We had an English canary called Manolito. He sang so loudly that at lunchtime we had to cover his cage with a cloth—it was the only way you could hear yourself talk. I also kept exotic birds such as Speke's weavers, cut-throat finches, and an American northern cardinal (which was probably my favorite). My father built a large aviary in the orchard of our *finca*—Spanish for a rural or agricultural piece of land, usually with a cottage. And from then on the list of birds and pets grew and grew. In the end I had so many pets that my parents imposed a ban on anymore feathery or furry family members.

A man named Félix Rodríguez de la Fuente was my hero. He was a medical doctor and was into hunting and rural pursuits, so he understood nature—to catch something, you have to understand the ways of your prey. In the late 1950s and early 1960s, he and a few English people resurrected the art of falconry in Spain, mostly using medieval books. In 1975, he became the presenter of a natural history program called *El Hombre y la Tierra* (*The Man and the Earth*). In this show he would do things like raise a goshawk, and also find a pair that was nesting, then combine shots of the trained bird with sequences of those in the wild, so viewers could compare how these birds lived and caught their prey. He also raised a pack of wolf cubs and trained them as hunting dogs, so if he wanted to film a pack of wolves chasing a deer he could release them and accurately predict where they would go, sometimes over several miles, positioning cameras along the route to film them as they ran. In much the same style as David Attenborough, he was an incredible narrator—intense, dramatic, and almost poetic at times. The funky psychedelic music that accompanied the images only intensified my love for the show.

My mother said that from as young as two years old, I would be asleep in bed, but as soon as the show's music started I would come crawling from my room and sit on the floor in front of the television, transfixed. I was a fan for many years. When people asked my friends, brothers, and sister what they wanted to be when they grew up, their answers ranged from a football player to a bullfighter to a fireman—sometimes even a general. I didn't know the word to describe the job Félix Rodríguez de la Fuente did, but I wanted to be just like him (and I still do).

Some of the images he created were unbeatable. I don't think that even Attenborough and the BBC had produced anything like them at that time. He mixed poetic descriptions and surrealism with scientific information. Because he knew animal psychology, he could anticipate what they would do. He would put carrion in the flight path of an eagle, with cameras focused on critical points, knowing that at some point the eagle would see the prey and pick it up and he could film what happened in detail.

I woke up on March 15, 1980, when I was almost eight years old, to hear the news that my hero had died in a plane crash in Alaska while filming a dog-sled race. I burst into tears. Without him, who would be left to so passionately plead for wildlife?

꘎꘎

I was about five years old when my father bought our *finca*. It was high on a mountain, about ten miles away from the outskirts of Gijón and twenty-five minutes from our apartment in the town. It was in the middle of a forest and, because most of the land was a peat bog, full of interesting plants. At the age of seven I identified a *Drosera rotundifolia*, or round-leaved sundew. It's a carnivorous plant with red, gland-tipped hairs on the upper surface and margins of the leaves, which sparkle like diamonds and lure insects to their death on the sticky tentacles. But my father drained the peat bog, because that's what people did: that way the land could be productive, as Franco had wanted.

It was not until years later, when I better understood the importance of ecology, that I realized the significance of what my father had done and the conflict that exists in certain places between the needs of people and nature. It was then that I discovered how best to transform a peat bog into usable land: by digging trenches for drainage. To increase the soil's productivity, we added truckloads of well-rotted cow manure, which we mixed in with the peat until it became a massive bed of multipurpose compost two or three soccer fields long and several feet deep. A strip was left intact, where the original plants survived.

In another area of the *finca* there was a hill of clay and conglomerate rock, so my father borrowed a backhoe and flattened it out. After a while he put an eight-inch layer of manure over that too, and the hill was transformed into a flat, flowery meadow.

My parents started to grow crops to feed the family, and my mother began gardening as well. I don't think anyone had much idea of what "organic" food was back then, but my mum certainly distrusted mass-produced food. She noticed that the eggs from the supermarket were starting to lose their taste, and thought they probably contained toxins, so she bought some hens and fed them on harvested leaves and vegetable waste— we would recycle everything. Corn for the hens attracted lots of mice, so she would kill them and the hens would eat them too. I soon came to see that chickens and dinosaurs had a lot in common.

At the *finca,* there was no television and no books, so working the land became your focus, your entertainment, your friend. We went there once a week to work, and in the end we had 2,000 fruit trees on the plot, including peach trees, apple trees, and kiwi plants.

It was here that I first grafted and budded a fruit. When I was about ten years old I had grown a kiwi plant from seed. One day I was chatting to a local nursery owner who sold his plants at nearby fairs, and I mentioned that it never flowered.

His response was sharp: "You silly boy. If you grow them from seed they will take several years to flower and you won't know until they do if they are male or female—the fruit may not be that good either."

Instead he showed me how to graft. Grafting is horticultural sleight of hand. Say, for example, a plant produces desirable fruit but its roots are weak. What a gardener would do is take the root system from another plant (called the rootstock) and graft a young shoot from the desirable variety (the scion) onto that, so that the desirable plant can grow on the "borrowed" stronger roots. The operation requires sharp tools, clean cuts, and skilled carpentry so the joints fit snugly together. Once this is done, you bind it to stop it from drying out and wait for the cuts to heal. It can be done in a couple minutes or less, with a bit of practice. If the scion shrivels, it has failed . . . but with skill you will have a new plant.

I did about four or five; one or two failed, but the others were successful. Grafting is the equivalent of an organ transplant in the plant world, of creating a Frankenstein plant. You need to follow a series of steps, have dexterity with a sharp knife, and be able to make a series of cuts in the right order. Speed and precision are paramount. Child-sized hands are good for that kind of thing.

The nursery owner would often sell the rootstock separate from the scion for the customers to graft themselves, with the comment: "If ten-year-old Carlos can do it, you can too."

My mother was the gardener. She would buy rare, beautiful, or tasty trees to plant. She might spend the whole day digging potatoes. She never stopped, planting more than I had ever seen: making a trench, then planting, making another trench, then planting. By the time I was six or seven years old I would help her.

We would stop at the market on the way to buy baby onions or Padrón peppers for planting. She even planted in the rain. We worked hard, but it didn't seem like it. I enjoyed gardening and eating; it didn't feel like forced labor. We had a rotavator, chainsaws (which I was happily wielding from the age of ten to twelve), and a small tractor and trailer.

One of the fears my mother had when they first bought the *finca* was that I would drown in the pond—even at an early age I was attracted to its water lilies and fish—or that I would go missing. She would wake up from a siesta thinking I was there, but I had gone. I quite often got lost (more in time than in space) in the surrounding forest while wandering about to see what I could find, being led farther and farther in by the sounds of birds, a distant clump of ferns, or simple curiosity.

I would frequently wreak havoc by bringing wildlife home. Once I adopted a badly injured, rather feisty gannet that I found on the beach. When a bulk carrier polluted the bay after an accident, I was allowed to keep a puffin or two. Wildlife hospitals did not exist, so I guess I was trying to fill the gap. With the help of my mum, we saved quite a few birds and animals, and those that did not die would eventually be "evicted"—set free from our apartment to go back to the wild. The apartment was so cramped because of the number of additional occupants that it finally got to the point that everything had to go, either flying from the window if it had recovered and was native or rehomed in the aviary at the *finca* or some distant zoo.

One day on the beach I came across a leatherback turtle weighing nearly half a ton. Luckily for my family it was dead already. It was taken away by biologists, who found that it had once been tagged in Venezuela. This blew my mind. I learned that this type of turtle has existed for more than 100 million years, surviving a few planetary-scale extinction events (mass extinctions), and that they feed mostly on jellyfish, are one of the deepest-diving air-breathing marine animals (diving to a half mile), and keep warm in cold water by staying on the move. That is how they end up on a rather chilly beach in northern Spain.

Throughout my childhood, when my mother spotted a plant, she would tell me all about it: the name of the species, where it came from, what it could be used for, and where she had seen it before. Gradually I learned the names, though I couldn't always find them in the encyclopedia because she used the local slang (and Franco never acknowledged the existence of other languages in Spain, even banning people from using them). Sometimes my father would have to stop the car so she could show me a specimen; sometimes we would collect seeds or cuttings.

When we left Asturias to go to a wedding, say, we would come back with vast collections of new seeds and specimens to plant at the *finca*. Later, when my father became a salesman for a company that sold gardening accessories and imported houseplants from Holland, there were always samples to spare.

The *finca* became our private botanic garden—though with the animals perhaps it should be described as a zoological garden.

As well as keeping an eye out for plants, my mother was always birdwatching. I started too, at age fourteen, though the term did not exist in Spanish at the time. It was unknown as a hobby. Shooting them, yes; watching them, no. I remember bumping into a friend of my family's one day at a train station at eight o'clock in the morning. He wanted to know where a fourteen-year-old could possibly be going at that time of day. I told him I was going to see the birds.

"What are you going to do? Are you going to shoot them?"

"No, I am going to watch them."

"Oh, that is odd. What do you get out of that, then?"

I told him that they were really beautiful, that they came from different countries to stay over winter, and that I would see many things but wouldn't know exactly what until I got there. The look on his face said it all.

There were lots of garden birds in the village during the nesting season; my mother would find nests and show them to me. Once, when I was very small, about four or five, she climbed with me to the top of a small pine to look at a goldfinch nest, using the alternating branches as steps. She could recognize many types of birds and knew their arrival dates. I remember her hearing the first cuckoo one year and telling me how it started life as an egg in another bird's nest, and how to identify it from its call and flight. She told me about the golden orioles, which chose the forks of the alder and poplar trees by the river to make their nests—perfect baskets of carefully selected, interwoven straw and poplar fluff. After they had finished breeding one year, she cut a nest down and brought it home to show me.

My godfather, Paco, observed the decline of species like quail, while he was scything the grass in the fields in spring. He would come across nests, unfortunately killing many of the birds because they hid by staying still, using their feathers as camouflage, so he wouldn't see them until it was too late. He would recite sayings that were related to the wildlife calendar, such as how on March 21 three bird species would cross the sea (swallows, cuckoos, and quail).

One day, when I was about ten, after I had been away on a trip to central Spain with my parents, I asked Paco why I had seen so many storks there but none in Asturias. "Oh, but there are," he said. "Some always nest in a bar in a village called La Espina, at the top of a hill, on their way farther south." When I was fourteen, I was at a meeting of ornithologists and they said the stork had been extinct in Asturias since the 1960s. I coolly told them what Paco had once told me. They went to the bar and asked the owners, who said they were there until the 1970s, and even had pictures of the nests. It was one of the last records of them nesting in Asturias. I later discovered that storks are gliding birds that need thermal currents to travel long distances. So they rarely stray from continental climates and avoid Asturias and Atlantic Spain.

❧ ❦

I felt frustrated at school, especially in my early years. It was old-fashioned and draconian. The system was based on your capacity to memorize; no one ever assessed your ability to understand what you were learning or encouraged you to do so. If you did, you might question or challenge the teachers. They certainly didn't understand the value of lateral thinking or creativity. The headmaster of the primary school was obsessed with *El Cantar de Mio Cid* (*The Poem of the Cid*), a medieval epic poem about El Cid, the famous Castilian military leader— he would recite passages from it. Even at age five you were expected to know the words or expect punishment.

From thirteen to seventeen, I attended a Catholic school run by left-wing Basque nuns who were part of the liberation theology movement—a movement that believed that the Church should help the oppressed. It was in stark contrast to my previous school. One day, one of my history teachers, who later became a member of Parliament in the Communist Party in Spain, was talking about the Second World War. He noticed I was daydreaming in class and said, "Carlos, what are you looking at? You are the only one in the class who has the personality traits of an Auschwitz survivor."

I was caught off-guard. "What?" I said.

He asked the class, "What do you think it is that makes Carlos like an Auschwitz survivor? Is it because he is physically stronger? No, because after a few months with no food, no one is strong. Is it about having strength of values? No, because they would be crushed. It is because whatever happens, however horrific, after three hours it becomes boring and they switch off. Their minds drift off to another plane, their imaginations detach them completely from the harsh realities they are facing, giving them mental breaks, so they are refreshed, revived, and don't go mad."

While my circumstances wouldn't have qualified as horrific, he was right that I was bored. My mind would take me back to my mother's village, to a place in the mountains or the aquarium at home, to the people, places, and subjects that I loved. I would spend weeks at a time researching dinosaurs, cloud formations, birds, minerals, fish. Often, though, it was plants I was interested in; I always seemed to return to plants.

My teachers made me do one of those ink tests—the Rorschach test, where they put ink on the paper, fold it in four ways to make a blot, then ask you to describe what you see. Some people see one thing, others two or even three—a woman's face, a cloud, the coast of Australia.

I described more than sixty things.

This seemed to indicate I had quite an imagination. The teachers wondered if it could be pathological, but it gave me an advantage: my mind could look at problem solving from many angles, so I had many more options than most.

By the time I was eighteen, I was just about fed up with the education system. What was the point of gaining a higher qualification in a country where more than 4 million were unemployed? If you trained as a doctor or a lawyer, you would be fine, but as a natural scientist? It was pointless. I loved nature, but everyone told me it would be impossible to turn my passion into a profession.

Though I had been birdwatching since I was fourteen and knew some people who were biologists, it was one of the most difficult degrees to obtain and carried with it the highest level of unemployment. It was a moment of despair. For the sake of having something to do, I decided to run a bar with a friend. While my father did not fully approve, he realized that I was young enough to learn and recover from a possible failure. Our bar, El Café de las Letras, served coffee by day and alcohol by night. It became a meeting place for intellectuals and philosophers, somewhere people would sit, smoke, and discuss politics and the arts. We had live jazz bands and music in the evenings, and we started to build up a good reputation. On Friday and Saturday nights there would be seventy people inside, with masses spilling out onto the pavement.

After a few years we sold it and paid off our debts, and I left that world behind. I was twenty-five. A year later my father grew sick with heart problems. We were told he could die at any time. It was only then that I made a concerted effort to begin a career in conservation, starting with a series of short-term jobs.

One of these was to observe the oystercatchers at the coast and deter beachgoers from getting too close to them. There were only about eight pairs of oystercatchers in Asturias (now there are up to twelve), and they nested on tiny rock islets, close to public beaches, which were accessible at low tide. As well as keeping away beachgoers, I observed everything that happened to the birds: when the first egg was laid; if the seagulls came and attacked them; when they flew off to feed. I had to log every moment of their lives, so we could learn more about them. It taught me to look carefully and record every detail. I was there with my telescope at five o'clock, before the sun rose, and would sit like Buddha waiting to be illuminated under his fig tree, until the sun set. I would watch the tide go out, come in, and then go out again, five days a week for the whole mating season.

My next job was concerned with Agenda 21, an international edict to increase wildlife numbers in towns and to take on projects to improve the landscape, environmental health, noise pollution, and wetland restoration. I was hired by the council to come up with some ideas of how they should do this. Over the years, they took a few things on board but not everything. They eventually created a public path in the east of Gijón, but the budget was small, which made life difficult and stopped me from doing everything that was really needed.

I also did some gardening, but it was poorly paid and people weren't really interested in anything other than general maintenance. I combined this with a bit of landscaping, and still worked in cafés and bars, including a trendy pub by the sea called Varsovia (Warsaw in Spanish), which people visited from all over Spain. I would start my shift at 1 a.m. and finish at 10 a.m. by going to the beach for a swim, temperatures permitting. I was a real free spirit. I met some fascinating characters, like the actress Amparo Larrañaga, the writer Arturo Pérez-Reverte, and the singer Javier Gurruchaga. One highlight was meeting Manu Chao, the half-French, half-Spanish singer who led a band called Mano Negra. It was a bohemian life, and gave me the confidence to mix with all kinds of people, including the depressed, high, and angry. But working in the bar never really made me happy; the only thing that could do that was wildlife, particularly plants.

❧❧

I was twenty-eight. My father had died, I had split with my girlfriend, and my last work contract had ended. Most of my ties had been broken. It was the ideal time to leave. My mother was sad—no one in the family had done anything like this— but realized there was nothing left for me in Asturias.

I saved some money and flew to Gatwick. I would try to speak the English I had learned in school, get a job, and, if it didn't work out, head off to another country. I didn't know what was going to happen, but I didn't want to return home. Failure was not an option.

My first job in England was as an assistant waiter at what is now the De Vere Selsdon Estate Hotel in Surrey. Within a year I became head sommelier, combining my love for horticulture with catering—"This wine is from this grape, which grows on chalky slopes, with a pH of 7.5," I'd say, happy to impart my knowledge. I'd learned a lot about wine from my father (my grandfather brewed cider—traditional Asturian stuff). If there were a hundred wines on the list, I could learn about them all (or most of them—if I told you in a strong Spanish accent about a grape variety, fermented in a type of oak, with a Latin name, in the cellar of an Italian lady, you would probably believe me more than you would a pale youth who looked like he'd never seen a vineyard). It helped to be a showman. I made an impact with the skills I had honed in my bartending days—five plates down an arm, that sort of thing. I would glide between the tables; people called me Nureyev.

It was a way of channeling nervous and creative energy. My mind never rests, so trying to find solutions kept me occupied and made me more efficient. There is always a better, easier, or faster way.

There was a historic garden and golf course at the hotel, and it didn't take long before I was creating flower arrangements for specially themed evenings or Christmas decorations. One year I contracted a stomach bug; I was banned from catering, so it was suggested that I work in the garden. I had to learn the English names of things—spade, shovel, pruning knife, mower—then the technical terms, including the different parts of a garden, such as terrace, avenue, and arbor.

On my days off, I would visit the sights in the capital—the Natural History Museum, London Zoo, the usual trail. Then, in November 2002, I took the tube down to the Royal Botanic Gardens at Kew. I instantly felt at home. I thought it would be good, but it was life-changing.

As I stepped through the giant iron doors into the Palm House, I glimpsed a mass of lush vegetation before being blinded as my glasses fogged up from the high temperature and humidity. The echo of the slamming doors made me feel like I was in some kind of plant cathedral. Then I was hit with that rich organic smell, the essence of a rainforest. I instinctively knew what it was, though I had never visited one before.

I realized that I was in one of the most biodiverse places on earth, even if it was artificially created. All the plants had labels that told you what they were and where they were from. Unlike in museums, the collections here were alive and well. This, combined with the beauty of the place and the fact that every corner was crammed with pieces of natural history, made it special.

Yet, still, I felt like I could make a difference. I spotted a few tropical plants in shade that should have been in sun—so there was room for improvement. I was not being critical; I just felt I could help. I decided this was where I should be. It's like when you fall in love: you can talk about beauty, style, personality, but in the end you feel that way for no reason other than love. I thought there might be jobs or courses available at Kew, so I decided to send in my CV. There just had to be some way I could get in.

On the tube train home, I found a discarded copy of a news-paper. Inside was an article headlined "The Living Dead." It was about how Kew was trying to save an extremely rare plant called *Ramosmania rodriguesi*. I was gripped. The writer explained that the plant, native only to the island of Rodrigues, had been feared extinct for the last forty years until it had been found by chance by a schoolboy. Kew had succeeded in growing on cuttings from it, which were producing flowers, but these flowers weren't producing seeds. Seeds were the only way you could ensure its long-term survival in the wild. The plant was beautiful and its history fascinating, but the prospects for its future were depressing.

"I've never seen this plant," I thought. "I'll have to go back."

I figured that there had to be a way for me to sneak into Kew through the back door. Looking online, I found there was a School of Horticulture. I emailed the principal and asked to meet him so I could explore my options and also, perhaps, see the plant.

Luck was on my side, and in December 2002 the principal, Ian Leese, kindly invited me to Kew to meet him. He read my CV and said all the things I didn't want to hear: "We have so many applicants . . . many are very skilled . . . you do not have that much professional experience . . ."

I decided to be bold. I was not going to let this chance slip away.

"Listen, Mr. Leese, I know that on paper my CV doesn't

sound that great, but I know something that is not written down. I know that I need this place, and this place needs me too. It is simple, if you ask me. You tell me what it takes to get a job here, and I'll go and do it."

Ian laughed, but the expression on his face seemed to say: *That's an interesting and novel approach to things.*

He thought for a moment, then said: "Okay, if you are as good as you say you are, there is a way you can prove it to us. It is called an internship. You would work here as if you were a staff member, but you wouldn't be paid. If you are good, and the essential asset you claim to be, sooner or later they will give you a job. Do you have any savings to support yourself while doing it?"

Luckily, I had. Now there was no going back.

"Which part of the gardens would you like to work in?" he asked.

"Any—but I have had a glimpse of the Tropical Nursery over the fence, and from what I could see, it looks amazing."

"Excellent," he said. "We don't have many people wanting to work with tropicals. A lot of senior management are based there, so you will be quite conspicuous too."

The deal was done.

2

Kew Calling

In Spain, January 6 is the main Christmas festival—the Feast of the Epiphany—and marks the arrival of the three wise men in Bethlehem. It is a day for the sharing of gifts. On January 6, 2003, I started as an intern in Kew's Tropical Nursery. It was the best Christmas present I ever had.

I remember the first day perfectly. It was nonstop, high-octane thrills—or that's the way it felt to me, at least. After a short induction, I was assigned to the aroid zone. Aroids are a group of plants that have a central spike of flowers surrounded by a "cowl," or hood. Anthuriums are one example of an aroid, as is the most famous of them all, *Amorphophallus titanum*—the titan arum or corpse flower—which we grow at Kew.

And guess what? A scientist who was working on the breeding system of *Ramosmania rodriguesi*—the critically endangered plant featured in the copy of the newspaper I had picked up on the train—had asked to keep six of these plants in the aroid zone so that they were separated from the other specimens. Even though they were not aroids, I would be looking after these plants too. It felt like a kind of destiny.

I was nervous. All the plants in the collection—about 500 species—had to be watered manually with a hose and a lance. Every species has a different requirement, and you have to assess how much each individual plant needs. A recently repotted plant drinks less than a plant that has not been repotted for a while, because the root-to-compost ratio is different. There were lots of decisions to be made, quickly and correctly, quite early in the morning. I felt better when I thought, "This is like running a bar or being a sommelier, but instead of providing drinks for people, you are doing it for plants." I had to apply what I already knew about watering plants, but on a scale I had never before attempted. The sheer size of the plants, their number, variety, diversity, and the various compost types made the challenge almost overwhelming.

The internship lasted three months. I was expected to learn as much as I could about the plants that I saw, particularly those I was working with—how to propagate each of them, how to treat sick specimens—as well as learn Kew's techniques and procedures. It was a task so vast it felt (and still feels) endless. There are twenty-one greenhouses in the Tropical Nursery, with about 44,000 plants; even after a month I felt I had glimpsed only the edges of the collection. But I was spoiled for choice when it came to learning from other people's skills: there were about twenty experts there, who I gradually got to know, plus contractors, visitors from other departments, interns, volunteers, associates, and researchers. "Plants, people, possibilities" was Kew's motto at the time, and a perfect summary of what the place is all about.

The greatest responsibility of all was looking after the plants with a red dot on their labels, indicating they were endangered. I had to collect seeds from these plants regularly for storage in the seed bank, as they were so rare in cultivation. It was vital they survived. This was not the kind of collection you would find in any garden; these were rare and precious plants. Gathering their seeds was a repetitive task, but invaluable because it allowed me to get to know each plant and its individual needs.

Just before I finished my time as an intern, I had a lucky break when a temporary position was advertised for a plant propagator at the Temperate and Arboretum Nursery. I went for it and got the job. Even though this role was a big step up, some at Kew noted my confidence. Perhaps they realized I had a bit of an eye, even a "sixth sense" for propagation. Thanks to my mum's collection of plants and her instructions, I nearly always had some experience with a relative of the plant I was dealing with, be it a fern, a bromeliad, an aroid, or a conifer. The job was nursery-based, but I was propagating all sorts of things: trees from the Palm House, shrubs from the Temperate House, and hardy plants from literally anywhere in the gardens. Struggling, sickly, or poor specimens were repropagated too, either by taking cuttings or growing from seed. Much depended on the species, time of year, quantities needed, state of the specimen, and the volume of material available. It was essential to have a plan for each propagation request (made by different managers of different areas). So often I would go out with Noelia Alvarez, the team leader, usually in an electric buggy, early in the morning to collect the plant material. Our conversations would be similar to those on a doctors' rounds in a hospital: we would discuss our patients, their condition, and what we could do to help them. If the plant was on public display we would have to select material carefully so we did not spoil its appearance. Occasionally there would be a special request. Once, Buckingham Palace wanted a rare mulberry tree, *Morus cathayana,* to add to their collection of mulberries (which are the only source of food for silkworms).

I was lucky: Kew had so many endangered plants in the collection, including the café marron, that I could keep interacting with all these species even when I had moved on from the Tropical Nursery. Nearly every day I would notice new shoots pushing out from the surface of the compost or roots peeping out from the bottom of the pots in the mist unit, a sure sign that cuttings had rooted. The sheer diversity of plants I was working with and propagating was daunting, but pleasantly so.

Applications for the Kew Diploma in Horticulture opened and I was lucky enough to be selected for an interview. Although there were thirty applicants who met the criteria, there were places for only twelve to fourteen students.

The Kew Diploma is one of the most prestigious horticultural qualifications in the world. The three-year course combines practical experience and theoretical study and is taught at the very highest standard. You spend time working in the gardens in nine different locations, growing alpine plants, climbing trees, and working with plants from the humid and the dry tropics. You have a block of lectures covering the sciences, landscape design, and plant identification. It is like studying at Oxbridge, Harvard, Leiden, or Universitat de Barcelona. It pushes you to the limit, and though nobody would do it twice, the long-term rewards are immense.

The year I applied, the BBC was filming a series called *A Year at Kew* and they wanted to film the selection process. As I crossed at the traffic lights on the way to the assessment, the BBC was filming me in the street. So as well as having to perform the various tasks, I was also introduced to media pressure. It was going to be a long day.

My CV was not that impressive, but the fact that I was already working in the gardens meant that the selection panel knew what I could do, what they were getting.

First was a plant-identification test. I was shown into a small greenhouse with thirty numbered plant samples. I had to identify them all, giving their genus, species, family (if known), and common name. Some were common garden plants, others less familiar. As I studied each plant carefully, I realized that the common ones were the trickiest, because you never use the family or Latin name. I trusted my gut instinct and tried to stay calm—not easy when the result meant so much.

We moved on to a random plant on a bench, sitting next to a selection of cutting tools, pots of different sizes, and several options to encourage rooting, including a mist bench and a tray of compost.

"Can you propagate this plant?" one member of the selection panel asked.

"Sure!" I said, grabbing a knife. Immediately, the questioning started up again.

"Why the knife, and not the scalpel or secateurs?"

They wanted to know my thought processes, not just my knowledge of the plants. I kept things simple. My feeling was that underplaying an answer was better than brashly responding as if I knew everything already. "I am not sure, but I think it's because secateurs damage the stem when you close the blades to make the cut," I said. "You want a clean cut that slices through the tissue like a surgeon's blade. Scalpels are fine for soft growth, but this stem is a bit woody, so a knife is the right tool to use here."

Finally I faced the interview panel, made up of senior members of staff, including heads of departments and senior horticulturists. They sat behind a long bench and fired off questions.

"Look out of the window. Can you see that tree? What is it?"

"It looks like *Pinus wallichiana*."

"Can you name five species of pine?"

"*Pinus nigra, Pinus pinea, Pinus* this, *Pinus* that . . ."

"You really know your conifers, don't you?" said another. "Name a conifer from the southern hemisphere."

"Erm, *Araucaria araucana,* the monkey puzzle tree."

It seemed to go on forever. The inquisition ended with one final question.

"What is your favorite plant?"

This was it—the point when I had to prove the passion I had for plants by demonstrating a deep knowledge of the subject. "*Ramosmania rodriguesi,*" I replied without hesitation.

I left the room proud to have survived but unsure of whether I had got through. A few days later, a letter arrived in a small brown envelope marked "On Her Majesty's Service." Against the odds, I had been offered a place. It was like winning the lottery of the horticultural world.

❋❋

And so, at the start of September 2003, I began a three-year course of epic magnitude. There were fourteen of us, ranging in age from eighteen to forty-two—English, Japanese, German, Korean, Irish, and Spanish; architects that had worked on skyscrapers in New York, Japanese ex-bankers, and hyperactive young English talents. We were an international circus troupe with a passion for horticulture and plants, plants, plants. This would be three years of competition and collaboration, pressure and parties, and deadlines, deadlines, deadlines.

Every two weeks there was a plant-identification test. Thirteen days before the test you would be told where in the gardens the plants would be chosen from, or the subject of the general theme. It could be forty grasses, forty trees for autumn color, forty different orchid genera, or forty plants from one particular tropical family. Then, on the day of the test, you had to name twenty of the plants from just a sprig in a flowerpot, with or without flowers: the genus, species, common name, family, and country of origin. Everything. The day after, you would find out the next group to be studied.

One of the requirements of the course was that we moved around the gardens in placements. I worked with orchids and alpines, then in the Princess of Wales Conservatory, the Rock Garden, and the Duke's Garden (a traditional English garden), then with the tree gang (climbing trees, carrying out tree surgery, and wielding a chainsaw ninety-eight feet from the ground), then in the Mediterranean Garden, and finally in the Conservation Area. I had never seen so many plants before. There was so much to learn.

Working in the Arboricultural Department enabled me to view Kew at every level, right up to the treetops. Hanging from one by a rope was initially daunting but ended up as a real thrill. I remember sitting on a side branch of a tall *Sequoiadendron giganteum* (giant redwood) overhanging the lake, early one morning, which was a perfect place to rest. There were tough times too: 8 a.m. in the Rock Garden, at 26°F, picking up pine needles individually from a cushion plant. I don't usually mind the cold, but this kind of painstaking work at freezing temperatures found me counting every minute until the next tea break.

Each year we had three lecture modules lasting three months: plant physiology, pathology, genetics, systematics, land surveying, landscape design . . . if it was related to plants and gardens, it was in there. Then there were exams and written projects, including studies on propagation, management, and classification. We went on trips to Wales, Cornwall, and Spain, and, if we were lucky, had the chance to study abroad in an exotic place—Australia, the USA, Congo, or in my case Mauritius. In the precious time in-between, we had to give at least one public lecture *and* submit a 10,000-word dissertation. It changed my life. You discover things you didn't know about yourself, the kinds of revelations that occur only when you are pushed to the limit and you still come out of it loving gardens and plants.

Micromorphology was fascinating: it's not every day you learn about plant anatomy at a microscopic level and discover the details of structures that can't be seen with the naked eye. That's a hidden world, and the secrets it revealed led me to a far deeper understanding of how a plant functions. Interestingly, each woody species has a unique "landscape" when you view it on a microscopic scale, so you can identify different types of wood from slivers or tiny fragments. Genetics was fascinating too, as was learning about plant breeding systems.

One day I'd be in the lab looking at microscope slides of the wood of an English oak (*Quercus robur*); the next day I'd be shaking with terror halfway up the chestnut-leaved oak (*Quercus castaneifolia*) near the Waterlily House. This tree, although not the tallest in Kew (that honor goes to a giant redwood), is thought to be the bulkiest in the UK. The base of the trunk is humongous, and as you start climbing, it branches into multiple trunks that get thinner and thinner. By the time I reached a thickness where I could almost hug it, the boughs were rocking in the wind and I could hear wood creaking. I just hung there, trying to comfort myself by musing on the tracheids, ray cells, and lignin—which I had seen on microscope slides—that ensure the trunk won't snap.

My mentor throughout was Ian Leese, head of the School of Horticulture. Late one night, as he opened the door to the computer room to switch off the lights, he saw me. "Oh, you still here, Carlos? *Buenas noches,*" he said, before heading off to collect his bike. I stayed until 2 a.m. Then, at 6 a.m., the ring of my mobile phone dragged me from my bed. On the other end was a distressed fellow student, who broke the news that Ian had died overnight. I was stunned. Any time I felt overwhelmed he would say: "It is simple, just keep going and you will achieve your goal." I often hear his voice in my head, even now.

❖

The Kew Diploma is an asset. After all, it is not only plants that are endangered—botanical horticulturists are too. Botanical horticulture is different from horticulture in general because you don't grow what you want in order to make a pretty garden, you grow what you need in order to maintain a collection.

It may be a plant brought from abroad by a plant collector or botanist that is new to science. You don't know the species or family, but you have records of where it comes from—the habitat, climate, rainfall, and specific location—to help you understand how to grow it. You have to be prepared to experiment, take a gamble, and, if necessary, work some magic and use unorthodox methods.

Does it have to be sown wet or dry? If you don't know, logic is required. If it can't be dried, and you try to dry it, you will kill it. Perhaps you have only a few seeds, with little margin for error; or maybe you have many, so you can carry out trials from day one. It may be from an equatorial location with constant temperatures or a place with temperature fluctuations, or it may need to pass through the gut of an animal before germinating. It may be from a species that grows in the branches of trees, or in water, or comes from high altitude or desert.

You have to analyze every bit of evidence, put the pieces of the puzzle together, and come up with something. It is like cooking by creating a recipe as you go: you may have to add an ingredient, change your frying pan, cater for a new guest, adjust the seasoning. The method for cooking porcini mushrooms in risotto is different depending on whether the mushrooms are fresh or dry. But there is always a science behind whatever you do; it *is* a science, even though it is performed with dirty hands and not in sterile conditions.

Anyone who claims to know everything automatically embarrasses themselves. Nobody knows half of this subject. In horticulture, the key is in what you don't know—and how that knowledge can be grown.

We learned a little history during the course too.

Plant conservation at Kew began in 1759, when Princess Augusta and Lord Bute created a botanic garden in the royal pleasure grounds. What started as a small collection was expanded by Sir Joseph Banks and King George III as the British Empire grew and more plants arrived. It gradually took over the whole of the gardens to create the vast collection we know today.

Banks not only collected plants himself, notably when traveling with Captain Cook on HMS *Endeavour*'s voyage to the South Pacific (1768–71), but began the tradition of sending out plant hunters. The first was Francis Masson, who brought back the iconic *Encephalartos altensteinii*—"the oldest pot plant in the world"—which still grows in the Palm House at Kew. The main reason for collecting plants back then was commercial, and to assist the expansion of the British Empire, but some plants that were taken to Kew and its satellite gardens in St. Vincent, Singapore, and Calcutta were already rare or became extinct in the wild soon after.

Subsequent regimes at Kew maintained and added to the collections, increasing their conservation value. Many of the earliest plants that were gathered are now listed in the IUCN (International Union for Conservation of Nature) categories "rare" and "critically endangered." Professor Jack Heslop-Harrison was the first director to initiate conservation programs such as seed saving, and work on Red Data books, which record details of endangered plants. From then on, the focus on conservation increased with each new director, particularly Sir Ghillean T. Prance, who oversaw the gardens from 1988 to 1999 and was a specialist in Amazonian botany. Conservation work grew in all areas, notably with the Millennium Seed Bank in 2000, which now holds just over 10 percent of all seed plant species and aims to contain 25 percent of the world's plant species by 2020.

❧❦

Kew houses the largest and most diverse botanical and mycological (fungi) collections in the world. This includes about 7 million dried plant specimens in the herbarium; a living collection of more than 19,000 plant species across the gardens and at Wakehurst Place; 1.25 million dried fungal specimens in the fungarium; more than 150,000 glass slides showing minute detail of plant micro-traits; 95,000 economic botany specimens, which show the extent of human use of plants; the world's largest wild plant DNA and tissue bank (including 50,000 DNA samples representing more than 35,000 species); and more than 2 billion seeds (from around 35,000 species) in the Millennium Seed Bank.

Conservation at Kew over the past three decades has balanced work in the gardens and programs overseas. It is a modification of the role that Kew staff once played when they moved plants around the empire, managing botanical gardens and commercial plantations like rubber or tea. Now they use their expertise in the field, teaching local people how to protect their own plants. Kew can do this vast work only because it is supported by myriad research programs in subjects like plant classification and molecular genetics, and has the backup of

laboratories, a large herbarium, a library, and masses of expertise from the people who work there.

It was a privilege to become part of that effort and to try to
continue the work that began when the first seeds were planted
in that patch of southwest London.

3

Resurrection in Rodrigues Island

Broad, glossy green leaves. Dozens of pretty little white flowers. The café marron plant—*Ramosmania rodriguesi*—from the island of Rodrigues in the Indian Ocean may not have won any gardening competitions, but it was pretty special. As I'd learned from a newspaper on my way home from that first trip to Kew, it was the last plant of its kind left in the world. Most experts had given up on the possibility of it ever producing the seed necessary for its survival.

Those experts, however, did not include me.

On the day I started at Kew as an intern in the Tropical Nursery there were six café marron plants waiting for me. I was amazed; it was an honor to look after a plant I valued so much. Someone from the Micropropagation Unit was working on them and had wanted to isolate them from the main group of plants. Day after day I saw them and gradually fell in love with the mass of beautiful flowers.

I began to read up on the history of this ill-fated plant. I first learned that islands are crucial for global plant diversity. They may make up only about 5 percent of the earth's land surface, but a quarter of all known "higher" plants (not mosses and lichens, but those with a "sap stream," such as woody and herbaceous plants) are the endemic or original native flora of islands. That's some 70,000 species. Diversity is usually higher on islands than it is in continental areas—which may come as a surprise—and plants are vital to the livelihoods, economies, well-being, and cultural identities of more than 600 million residents—just under one-tenth of the world's population.

The Mascarene archipelago of Mauritius, Réunion, and Rodrigues is a particularly special group of islands. Over many millions of years, they have evolved a unique and fascinating fauna and flora found nowhere else in the world. Yet the recent havoc wreaked on them by human impact—through land clearance and the introduction of alien plant species and animals—has led to an ecological disaster, dramatically reducing the number of native flora species. (This man-made "extinction event" has decimated the native fauna too: just a scarce few endangered species of land birds and fruit bats have survived the increased exposure to humankind, along with several species of reptiles and numerous invertebrates.)

The story of the café marron is all too familiar. Of the 1,296 native plant species on these islands, 53 are now extinct and 393 of the surviving species (just over 30 percent) are threatened. Little remains of the native ebony forest; eight of the ebony species are threatened. Six of their unique orchid species are extinct, and thirteen more are threatened. On Rodrigues, the smallest island in the group, at least eight species of "higher" plants are extinct. Of the thirty-eight surviving native species, twenty-one are endangered and at least ten of those survive in populations of fewer than twenty specimens (seven of the native species are known to have five or fewer specimens remaining). Less than 1 percent of the original habitat still exists.

Mauritius and Rodrigues have both been labeled "Islands of the Living Dead" because at least thirty of their species have stopped reproducing in the wild. With no prospect of producing future generations, the death of the last plant in each species signals its extinction. It is a horrific example of an ecological disaster, one of many.

François Leguat was an early settler on the tiny island of Rodrigues. In 1708 he wrote *A New Voyage to the East Indies,* in which he described being overwhelmed by the beauty of the island: "We could hardly take our eyes away from the little mountains of which the island entirely exists; they are so richly spread with great and tall trees . . . whose perpetual verdure is entirely charming." Destruction started not long after Leguat's settlement. In 1877, a mere 169 years after Leguat was so entranced, the botanist Sir Isaac Bayley Balfour observed: "It is difficult to recognize the barren and arid Rodrigues of the present day in 'the little Eden,' 'lovely isle,' 'earthly paradise' of Leguat. Fire, goats and finally introduced foreign plants well-nigh exterminated the indigenous flora . . . leaving the island, a field for the rank and rapid growth of common tropical weeds."

Unlike Madagascar, which split from the African landmass carrying plants and animals native to the continent, Rodrigues is a volcanic island, in the middle of the Indian Ocean, surrounded by a coral reef, about 350 miles east of Mauritius. It formed after a tectonic movement 10 million years ago, enabling unique flora and fauna to develop. As you fly to the island, you start glimpsing what can only be described as a smoke ring in the middle of nowhere—the bubble where Rodrigues Island "lives." After all, Rodrigues is more sea than it is land: a small territory with a private ocean almost sealed off by a coral reef barrier. The sea is usually calm inside the barrier, which is thirty to forty miles wide and protects the island; the color of the water is impossible to describe, as it is ever-changing, depending on the light and weather.

As you approach the island, you can see a few other inhabited islets. Rodrigues sits firmly in the middle of them. From afar, it looks like a single living cell in which the barrier is the permeable membrane. The islets are like cell organelles; the main island the nucleus.

Rodrigues itself is a small mountain range with a few flat areas. The airport is at a place called Anse Quitor (with all the mountains, it's the only place it could possibly be located). Landing is quite an experience: the instant you touch down, the pilot slams on the brakes and you are thrown violently forward in your seat until the plane rolls to a stop right at the end of the runway. Taking off feels like you are about to dive into the sea head first.

The island is still beautiful to the ecologically untrained eye, but to anyone with some idea of natural history, it quickly flags itself as a disaster zone. On my first visit, soon after leaving the plane, I saw vast expanses of highly invasive, nonnative *Lantana camara,* which is so aggressive in its growth that it suffocates benign native species. As hopeless as this looked when I first saw it, as of 2017 it has thankfully been declining at an extraordinary rate, mostly due to the introduction of the lantana lace bug (*Teleonemia scrupulosa*), whose larvae devour the leaves. Hopefully, other solutions will appear to deal with the other invasive species in the Mascarenes, for example, in the forests, with their impenetrable thickets of planted eucalyptus, guava, and the tree weed *Vachellia nilotica.*

Dotted with eucalyptus plantations, Rodrigues has a highly humanized landscape, almost like the Asturias where I grew up, with steep hills, roads, towns, and villages. I never expected to find such a place in the tropics. The native tortoises and an extinct flightless bird called the solitaire have been swapped for Western staples—chickens, goats, cows. A unique ecosystem has been destroyed and replaced with what you get throughout most of the world: monoculture forest and random farming.

A passion for conserving plants on the edge of extinction is infused within every molecule of my body. I believe that every species has a right to live without justifying its existence and should not be wiped out by recklessness or economic interest. We can't just pick and choose which plants to conserve, keeping only the ones that provide medicines or look pretty in gardens.

Destroy one species and you give yourself permission to destroy them all. We still know so little about what they are capable of. It is like finding a library where the books are written in Chinese, then taking someone to visit who can read only English and Spanish to decide which books are relevant. Or perhaps going into that library and burning the books based on whether you like the cover or not. My feeling is that all plants have not one but many uses. Perhaps they are not useful at the moment, but they may have been in the past and possibly will be again. We must not let them die out, for the sake of future generations who may need them to survive. Furthermore, when you lose one species, there are dire consequences for everything else that depends on it—insects, birds, and mammals, including humans. These are ecosystems, they need one another. I will not tolerate extinction.

In the 1980s, when people suddenly realized that mass extinctions were taking place, there was a global movement to conserve island floras, many of them "endemic" (the description given to plants or animals that are native or restricted to a particular place). There is a rule in conservation that decrees that if a species has not been seen for fifty years, it is declared extinct. By the 1980s, the nearby Mauritius had already witnessed the extinction of the dodo and thirty species that existed nowhere else in the world; in addition, sixty-one other species had also disappeared from the island. It was an ecological disaster.

On his visit to Rodrigues in 1877, Balfour discovered and named many new plants. He noted that one of them, *Randia heterophylla*, was "a remarkable heterophyllous species of a genus hitherto unknown in the Mascarene Islands" (*heterophyllous* means "having different kinds of leaves on the same plant"; the juvenile and mature leaves of these plants are distinctly different in shape). As years passed, the natural habitat of this precious plant was gradually eroded; it hung on in the outer reaches of the island for a few years before—according to the rule—it was finally declared extinct.

Then, in 1980, Raymond Ah-Kee, a teacher at a school in Rodrigues and a keen naturalist, set his class some homework. Wanting them to learn more about the natural history of their island, he instructed them to bring back some plants for identification, so they could talk about them in class. Most of his pupils returned clutching the rank tropical weeds that choked the local landscape. One student's collection, however, stood out. Hedley Manan collected his plants six to nine feet from the town's busy main road. As he had with the others, Raymond started to sift carefully through Hedley's collection, naming the plants one by one. He was familiar with most and identified others using a local flora reference book.

But one plant puzzled him greatly. He had never seen it before, nor was it in his books. It had white flowers, so he thought it had probably escaped from a local garden. He thought it unlikely to be an endemic species because of where Hedley had found it—what remained of Rodrigues's original species tended to hunker down in secret locations in the countryside, or clung on to survival in inaccessible places, such as cliffs plunging thousands of feet into the sea.

Failing to reach a satisfying conclusion, he carefully pressed the plant between two sheets of paper, put it in an envelope, and posted it to the Royal Botanic Gardens, Kew. It was forwarded to two scientists in the herbarium—Deva D. Tirvengadum and Bernard Verdcourt—who were experts in naming and classifying plants from the region. They compared it to Balfour's three pressed specimens of *Randia heterophylla*. Something didn't match. The original description showed that there were two different forms of the plant with the same name. The specimen with more flowers was most like the plant Mr. Ah-Kee had sent to Kew. The trouble was that there was no living population to compare it with, just the pressed specimens and a botanical illustration in black ink. The taxonomists could only speculate that either there was one extremely variable species on the island or there were two distinct species.

After Balfour had collected and pressed his specimens, he published a description of the plant in the *Botanical Journal of the Linnean Society,* giving it the name *Randia heterophylla.* This became the first priceless "type specimen" (there is only one, never a second), by which all others thought to be examples of that plant could be compared, identified, and in this case renamed. Of the 8 million specimens in the herbarium (the collection of pressed dried plant specimens at Kew), more than 350,000 are "type specimens," which are identified by the bright red line along the edge of the files they are stored in. There are specimens at Kew collected by Charles Darwin and David Livingstone, among others; each specimen is a piece of history, art, and science. Plants can't be identified and classified if there are no botanic gardens with herbaria and educated people with experience and knowledge to study and preserve these specimens for generations to come—this fact in itself justifies the existence of Kew Gardens.

Having looked at all of the genera in Rubiaceae—more than 9,000 species—and finding no other match for the plant Ah-Kee had provided, the scientists finally decided to create a new genus for the plant, calling it *Ramosmania* (after two of the first high-status officials on Mauritius: Sir Seewoosagur Ramgoolam and Sir Raman Osman). They designated the species *rodriguesi* and, like Balfour, published the change in a respected journal in 1992. *Ramosmania rodriguesi* was immediately classified as critically endangered.

I was surprised to hear that the café marron had been discovered by the side of a road, given the history of habitat destruction on the island; road building and forest clearance for agriculture and housing had long ago taken their toll. I had assumed the sample was found in a remote valley, where even the goats don't go. In Spain, botanists sometimes look for plants by the side of the road—there are occasional gems to be found there—but in Rodrigues? No. Hedley Manan's plant had probably never even been spotted before—though I've studied Balfour's pressed plants at Kew and often wondered if our material came from the same plant that he discovered and took his original specimens from.

Once the plant had been (re-)discovered in the 1980s, word spread rapidly. The world's media seized upon the café marron: it made the radio, television, and the front pages of newspapers. People were desperate to get to Rodrigues to see it. Then, early one morning, at the height of the media frenzy, conservationists went to visit the tree where the plant material had been found and were shocked to find that all that was left was a stump. It had been chopped down in the middle of the night.

The tree had been used in traditional medicine by the locals for years: some believed that if a child had nightmares, you gave it a teddy bear to sleep with, then threw the teddy at the tree.

(When I first visited, there were two or three teddy bears leaning against the cage that had been built to protect the plant.) It was also used as a treatment for venereal disease. Its fate was sealed, however, because of its use in an invigorating tonic to cure hangovers. (The warden in Rodrigues said to me that if I was offered money every time I was asked for a branch to make this tea, I would become very rich.)

As the authorities became more aware of the gravity of the situation, the plant was ring-fenced. Soon they realized that people would simply jump the fence. Another fence was added, this time enclosed at the top, so the plant was protected in a cage. After millions of years of freedom, the last plant in the wild had to be caged to protect it from humans. Even then, though, it wasn't safe: locals who knew of its whereabouts would start chopping pieces off, given half a chance.

The scale of the threat was clear. Wendy Strahm, the World Wildlife Fund project manager who worked on the conservation of the flora of Mauritius and Rodrigues, realized drastic action was needed. What was left of the last surviving tree needed protecting, and it was vital that material from the tree should be taken to a secure location so that research could begin on propagation. Increasing the number of plants would increase the chance of survival—crucial, because since the species had been rediscovered, it had never set seed.

The plan was to cut three branches from the newly re-sprouted stump and drive them to Rodrigues Island Airport, where a small aircraft that flew weekly to Mauritius would be waiting. The plant material would travel with the pilot to Mauritius and would then be transferred to a British Airways flight to Heathrow in London, again traveling in the cockpit of the plane. Once the precious cargo reached Kew, it would be split: some of the material would go to the Micropropagation Unit and the rest to the Temperate Nursery.

The plan was a success. The onus was now on Dave Cooke, senior propagator in the Temperate and Arboretum Nursery. He and a supervisor had already been working with people out in Mauritius, Réunion, and Rodrigues on the conservation of other rare species like the Mandrinette, *Hibiscus fragilis.*

There had been plenty of similar experimentation. Peter Tindley, who worked with Dave, was an expert on the propagation of "woody" plants like trees and shrubs. From 1984 to 1988, the Palm House at Kew was restored, and many of the plants in there had not been propagated before. The team came up with some ingenious techniques to propagate plants, like the very old mahogany trees, mostly based on experiments with different mixes of rooting hormones.

The café marron material was taken straight to Jim Keesing, the plant health inspector at Kew. From there it went to the nursery, where everyone became very excited about something that could be described only as "a stick with a couple of thin leaves on," which was about the length of a pencil (another piece was half this size and arrived wrapped in tissue and damp newspaper in a Jiffy bag). Dave and the micropropagation team were all looking at this larger piece of material on the potting bench when Dave thought, "I could root this." He promptly picked up his secateurs and cut it in half. There were gasps of horror and a few choice swear words.

"What do you think you're doing?" they asked.

"I chopped it in half to double my chances!" Dave replied.

At that time, the propagation facilities in the nursery were basic. There were heating cables overlaid with a mix of peat and sand, and polyethylene sheets draped over metal hoops covering the cuttings, creating 75 to 80 percent humidity. Dave took one of the cuttings and pared off a tiny sliver of bark from the side, just a few millimeters long. He left the other as it was. He dipped both their bases in liquid rooting hormone, pushed them into the rooting compost, and left them to see what would happen. Within a short time the smaller cutting had died, but several months later Dave noticed that the little buds on the larger one had started to open. He dug around in the compost with his hands and found some roots. Dave's frantic whoops brought everyone running, and the cutting was carefully potted and placed into a weaning case. The rest is history: by 1988, the plant had reached a reasonable size, so Dave took a cutting from the top of the main shoot, increasing the "captive" population by 100 percent. Another two or three cuttings were taken, then it was replanted in the newly restored Palm House and left to grow and flower.

It is extraordinary to think that Kew's café marron collection comes from a single branch. It is one of the only species that never stops blooming. In the fourteen years I have been at Kew, these plants have always been loaded with flowers. Aside from the one in the Palm House, there are two in the Princess of Wales Conservatory and eight other specimens, all from the same parent cutting.

The café marron was well established and flowered prolifically year after year before somebody asked, "Where are the seeds?" Confronted with this question, scientists started studying the stigmas (the surface where the pollen lands), the styles (part of the female sexual organ of the flower) and the pollination systems of related plants and came to the conclusion that there was nothing traditional techniques could do to encourage seed production. They considered other scientific methods too, but for various reasons these were rejected.

So, with plenty of specimens to experiment with, they began to make a list of possible barriers to propagation. The most important discovery was that it wasn't a genetic problem—the pollen was fertile and there were plenty of fully developed ovules (a plant's equivalent to human eggs) in the flower. The problem was with the pollen tube—a hollow tube that develops from each pollen grain and penetrates the style. In this case, the pollen tubes were growing a little but then stopped: there was a mechanism within the flower itself that blocked the growth of these tubes, stopping the sperm cells in the pollen from reaching the ovules.

In that moment, all the elation and hopes for the plant's survival were crushed. Sure, you could propagate indefinitely by using cuttings, but the plant could never produce the seed necessary to survive by itself. It had now joined the ranks of the living dead. Returning it to the wild would be pointless—about as useful as planting tulips in the middle of a roundabout. After twenty years of prolific flowering, it had never set a single fruit, nor was it likely to do so. Eleven plants had been reintroduced to Rodrigues, but they would now be an ever-blooming reminder of what had been lost, nothing more than hopeless cases sentenced to life imprisonment, captive in cages.

❦

Even in my earliest days at Kew, I just couldn't come to terms with the fact that this plant was "the living dead"; surely there had to be a way to make it produce seed. The problem lodged in the back of my mind and would not go away. I came up with all sorts of ideas. What if I grafted it onto another member of its family? Sometimes that leads to a stronger plant—would that change its ways? In commerce they often graft the style of a related variety onto the ovary, put the pollen in, and attempt germination.

Sometimes things that look complex are quite simple. One day it occurred to me that this plant never stopped blooming. There were masses of flowers opening every day, each lasting quite a long time. Each plant had twenty to thirty flowers—multiply that by six plants and I had 120 to 180 opportunities to try something different, to see if it could be propagated. I suggested this to those around me, but they looked at me as if I were crazy. The overwhelming response was "Don't waste your time," with the underlying hint that I should be doing something more productive. So I resolved to pursue it on my own time. I will not accept extinction.

I stayed on at the nursery late into the night, dissecting, analyzing, and observing the flowers. I began thinking about self-incompatibility systems in plants. One is where the stigma recognizes its own pollen and does not allow it to develop a pollen tube; the other is where the pollen tube is allowed to grow but is blocked before it reaches the ovary. So how do you overcome this? I didn't have a scientific answer, but I had a hunch.

I am allergic to grass pollen and every spring I am afflicted by hay fever. I'm all too aware that when I ride my bike and the grasses are in flower, my eyes get red and sore. The pollen lands in my eyes and starts to germinate and grow in the moist solution that is my tears. In fact, pollen can germinate on contact with distilled water—we use it as a standard method for germination. So I thought, "What if I cut off the stigma, somehow removing the lock, and put the pollen in the moist wound created by the seeping sap? Perhaps the pollen will find a way in."

Next I thought, "How long does it take to pick up a flower, take a scalpel, cut across the flower, and then put some pollen in the style?" I kept on doing it and doing it, but nothing happened. I knew, though, that if I did it again and again, more than 200 times, eventually the conditions might be right for a single grain of pollen to reach the ovule, pollinate, and produce seed. If I could grow just one—or, even better, a few—of these new, distinct individuals and cross-pollinate them with the last surviving clone, their offspring would restore fertility to the species.

※※

The summer of 2003 was one of the hottest ever recorded in the UK. There were days when it was like "Costa del Kew" inside the nursery, and I sweltered in temperatures over 100°F. I was on weekend duty during the August bank holiday, quietly feeding and pruning the plants in the shade, when I suddenly looked up at one of the café marrons.

There, right in front of me, was a fruit. It looked like a small green fig, about an inch long, pointing upward from one of the branches.

I couldn't believe it. There was a surge of blood to my head. I felt giddy; it was like scoring the winning goal in a World Cup final. I began cheering and dancing, then had to tell someone. I grabbed my phone and started calling people. I gabbled. People rushed into the nursery, urging me to calm down.

The seeds took about six months to ripen. The birth was slow and painful. I became agitated and was checking it every day to make sure the fruit had not been aborted. We sowed seeds from this first fruit in sterile conditions in the Micro-propagation Unit and they germinated, two tiny leaves appearing. Their quick death left everyone bitterly disappointed, but it proved that the seeds were viable. It was a positive but unsure step toward success.

With plants, obsession and passion are the key, otherwise you don't get anywhere. If you continue with traditional techniques, you will never push boundaries or make new discoveries. You have to become obsessed in order to progress. Yet time and time again with the café marron, there was conflict with the establishment. One person said to me, "You have done this a thousand times, and not a single fruit was produced. You are trying to tell us that this is a proven technique, just because you have one fruit now?" Others said that claiming outright success after producing one fruit from 180 flowers was just not professional. But I was not undertaking scientific research I was intending to publish. I just wanted to get a seed. Some scientists were almost inferring that I was putting them in an awkward position, as everyone would want to know how we had a fruit, but we couldn't say. My frustrated response was: "Yes, but I have solved the propagation problem and now these plants set seed. The problem of not knowing exactly how this happened is much smaller than 'We can't propagate this.' Now none of you can tell me that this is a waste of time." I had irrefutably found a way of producing seed of the café marron. It may have not been a controlled experiment, but it wasn't magic either.

To answer my critics, I needed to find out why this one plant had produced a fruit. What could have happened during the period between pollination and seed production? Several possibilities crossed my mind. One was the temperature, which had been hotter than ever—especially as the greenhouse shading in the place where the seed-producing plant was growing hadn't been working around the time the fruit was set. In the wild in Rodrigues, where temperatures are higher than they are in the UK, the plant grows in the shade. At Kew, though, the fruiting plant was the only one that had been grown in the sun, as far as I knew.

I returned to the scientific papers on the café marron. No one had studied its floral biology—how long a flower lasted or how long the pollen grains survived in the flower. I began to track the flowers from the moment they opened to the moment they fell, and found this took about seventeen to twenty days. From day eight the pollen went brown. "So why do the flowers stay there, looking nice, for two weeks?" I thought. "There's no need to attract pollinators if you are not producing pollen and you are a male. But then ... *are* you a male?"

Many flowers undergo a sex change—they are male for a time, then switch to being female, or the other way around—so I focused on trying to see if there were changes in the female parts. I noticed that while all the flowers started as male, on a few rare occasions the stigma—one of the female parts of the flower—which had been coiled, elongated a little and eventually opened up, a bit like a snake's tongue. However, by the time the stigma opened fully, the pollen was overmature and had degraded, so the plant was never in a position to self-pollinate naturally; it could be done only artificially with pollen from a new flower.

Most puzzling was why this occurred in some plants but not others. The plants in the Temperate Nursery were not doing it in summer or winter. The plants in the Tropical Nursery were doing it in the summer, but the plant in the Palm House, which I was keeping an eye on, was doing it in the winter. This did nothing to help my theory about the influence of sun and heat. What was going on?

I then realized that in the Palm House they would cut back the canopy in winter to increase the levels of light and sunshine reaching the plants down below and would crank up the heating on frosty nights.

Even though I wasn't working in the Palm House at the time, I decided to try to pollinate that plant. With the branches overhanging the heating pipes and in full sun, I transferred some pollen and one fruit appeared. I deduced that the recipe for success was sunlight, higher temperatures than we had thought, and the transfer of pollen from a fresh new flower to a mature one. But I still needed to prove it.

In the meantime, the plant in the Palm House was on public

display. If it were widely advertised that it bore fruit, visitors or plant collectors might steal the fruit, or an overenthusiastic gardener might remove the seeds. As the fruit takes six months to ripen, there was also the danger that one of the staff would forget and cut it off by accident.

One day, a supervisor was in the Palm House collecting samples for a plant-identification test for the students. In a rush, and wanting a good example of a selection of plants from the major tropical families, he probably just noticed the family name, Rubiaceae, and not the name of the plant. Snip! Off came a stem of the café marron, and with it, the fruit. Hours later, a member of staff came rushing in and said that a part of one of my precious plants was missing. I shrieked. I skulked into the student plant section, saw the shoot, and, on closer inspection, found the fruit hidden underwater, in the vase. Beheaded.

We sent it to the Micropropagation Unit, but once again I was fruitless. I also had to prove the point that heat was the key to advancing the flower into its female stage and increasing the possibility of fruit. So in the winter I took six of the plants in the Tropical Nursery and placed them close to the heating pipes, then opened up the shading system so the sun's rays could hit them.

Progress was slow—only one in about a hundred attempts at pollination were successful—but it felt like we were getting somewhere. After a while, eight fruits, containing five to eleven seeds each, were set. Since ants were stealing the nectar and moving between flowers, which I thought could contaminate my experiments, I decided to make an artificial island, a mini-Rodrigues, in a hidden corner of a pool in order to protect one specimen from them. I could check it when working with the waterlilies in the pool and record its progress from the day a bud emerged to when the fruits appeared. This mini-Rodrigues also ensured that the fruits would not be stolen or cut by a gardener by accident. Once upon a time there was one Rodrigues Island, with a single clone of café marron. Now we had two islands, and the second filled me with hope. My experiment was successful, and I succeeded in getting seed from the flowers that I manipulated.

Over the course of a year, 300 seeds were harvested from several plants, both in the Tropical Nursery and on the mini-Rodrigues. Finally the "seedless curse" appeared to be broken.

After the seeds were extracted from the first fruits, there was another dilemma: to dry or not to dry. Some seeds can't be dried at all. If you do, they will die. For others, the moisture content has to be dropped inside the seed before sowing. My gut feeling was that they would need to be dried, or at least tolerate it. If I was wrong, there were other fruits coming to maturity, so I knew there would be more chances to test things. After drying them in silica gel for a week, we sowed them in a specially formulated compost and waited. Seeds in this family can take a week to germinate—or a year or two. Patience is crucial, but you really don't know when is too early or too late. It is no surprise that I inspected the compost surface daily, sometimes twice.

Nothing happened for days, then all of a sudden my agony was ended. One seedling had already stretched up the stem, but the seed had not released the leaves. It looked like a safety match, standing and waiting to light up, to ignite a new life. On closer inspection there seemed to be a bit of raised compost, just over an inch away from the first seedling. I pursed my lips and blew gently; there was a second seedling pushing its way up. In the end they germinated in three weeks—the first seeds of this plant ever to be propagated in cultivation. The last flowering adult was seen in the 1940s, so no one could have seen a seedling for almost a century.

Some people still did not believe the good news. Initially

they had said that the fruit might not contain seeds, then that they wouldn't germinate. I was able to respond, "Here they are!"—we had both the seeds and the seedlings. Then, when they saw the seedlings, some said, "Perhaps you're cheating?" "Okay," I responded, "let's look at the DNA." That was a good idea, as other sceptics were telling me that I would need more than a single plant for genetic diversity. We examined the seedlings from the last wild café marron at a molecular level and found that, like the human race, the plant was very variable, so the parents of the last plant would have contained a mass of inherited genes, which could only be unlocked by seed production. The production and germination of seed was a kind of slap in the face to anyone who thought we were wasting our time.

In conservation, experts sometimes declare that there is not enough genetic viability to reestablish a species—and then give up on them because they are never going to go back into the wild. But even if it is down to a petri dish with a cluster of cells, I will keep going, keep trying. You never know what is going to happen. In my dissertation at Kew I set out to discover the minimum number of specimens that can be left before it is not worth doing anything to save a plant. The answer, I found, is zero.

I wanted to discover at what point inbreeding would cause a species to collapse, but then started asking questions to which nobody had the answers. For example, how come *Euryale ferox,* an Asian species of giant waterlily that has been "selfing" for millions of years, to the point that it doesn't even bother to open its flowers (and even when it does, they are already fertilized), is an invasive weed—one of the most vigorous plants you can find? Or look at the rabbits in Australia, which came from two female rabbits that were impregnated by the same male: in 200 years the size of the population has spiraled out of control and is now partially immune to myxomatosis. Doesn't this suggest there is more to genetics than we think? If an inbreeding "bottleneck" happens but is not lethal, it seems that even when a species is reduced to one it can be restored. The café marron from Rodrigues was once part of a huge population of trees and pollen. Its ancestors would have included hundreds of different trees over time. Even though it won't contain all the genes that were found in all of the variations, it will contain enough genes from the mother, father, and other ancestors down the generations for the species to be restored.

On a related note, how does it happen that the islands have flora too? Is anyone going to tell me that if a plant colonized an island, it did so in groups that were large enough to make the population viable? No, it is more likely that a single seed, or a few individuals, arrived on an island and found the opportunities so great and the pressures so few that the plant exploded in diversity, generating thousands of species. *Echium,* a genus of plants with many endemic species in the Canary Islands, all probably came from a single plant—on isolated islands, even genetic mistakes have a chance of survival. And yet there is always an excuse not to bother.

Because of the genetic exchange through sexual reproduction, the café marron plants we have grown are genetically variable. Importantly, we noticed quite quickly that they fall into two distinct types. Some bloom prolifically and have identical flowers to the self-pollinated parent. These produce viable pollen but no fruits—at least, not without serious encouragement. The others have flowers that are solitary rather than in groups, with a style that projects above the pollen-free anthers. When pollen is moved from the first type of flower to the second, they fruit freely, with each fruit containing an average of eighty seeds, much more than those that came from the original cutting. In other words, there are now both male and female plants. Most important, the species can now be reintroduced, and its populations restored.

I noted several other interesting things too. The leaves of the seedlings and juvenile plants are very different in appearance from those of an adult. It's a feature of plants in Rodrigues and Mauritius, and is an adaptation to protect them from grazing animals.

I know from my farming background that cows know what good pasture looks like from a distance, but close up they select what to eat using their sense of smell, while using their eyes to look out for predators. They will smell rather than see a poisonous plant, which is ideal when you are grazing thick grass or a flowery meadow. For plants in areas like this, visual camouflage is useless—but it *is* the only way to protect yourself from tortoises on Rodrigues and Mauritius, as they can spot lush foliage from some distance away. The café marron therefore camouflages its easy-to-reach juvenile foliage. The young leaves are thin and narrow, making them almost invisible to the tortoises, and are brown, silver, and red in color—they appear to be dead from afar. Even if spotted by a tortoise, the leaves would be difficult to eat, as long, narrow leaves are difficult to bite in comparison to broader leaves. When the plant reaches four feet, beyond the reach of a tortoise's neck, the leaves become elliptic in shape and dark green. A similar thing occurs in New Zealand: the foliage of the genus *Pseudopanax* changes at eleven to thirteen feet, just above the height of a moa's beak.

❧ ❧

In 2007 I received funding from a scholarship program to take fifteen small café marron saplings and some seeds back to Rodrigues. I rather liked the idea of Jose Carlos Magdalena

Rodriguez (my full name) going to Rodrigues—it felt like home. The response from the people on the island was overwhelming, as they had written the plant off. They were also keen to grow it as a garden plant, because it flowered constantly.

When we arrived, some of the plants were retained in quarantine on Mauritius as there was no facility for them in Rodrigues. The authorities were desperate to ensure no pests or diseases were introduced. I expected that to happen, so I brought 600 seeds as well.

All of the adult plants at Kew were grown in greenhouses and would start to go brown and look as if they were getting scorched when they were moved into the sun. People would freak out and move them back into the shade. This led to the conclusion that they grow only in shade. Wrong. On my first visit to Mauritius, I was amazed to find that one of the plants repatriated by Kew a decade earlier was flourishing in the full blast of the tropical sun against a white wall at the National Parks and Conservation Service's nursery. Its leaves were beautifully mottled in tones of brown. The ones in the sun at Kew were not getting scorched at all, they were just reacting to sunshine by producing the natural shades of brown that are normal in these conditions—they were getting a tan! I looked at the mass of flowers, realized they were fertile and perfect for pollination, so I pollinated them. When I returned about two weeks later, there were a couple of fruits forming on the male plant.

On my second visit to Rodrigues, in April 2010, I went on a pilgrimage to pay homage to the original tree, accompanied by Alfred Bégué from the Mauritian Wildlife Foundation, and a forestry guard. I could not believe what I saw. Rather than being treasured and pampered like royalty, the tree and its surroundings were in an appalling state. Since the species had been secured elsewhere, perhaps people had lost interest in preserving the last mature specimen left in the world; perhaps no one wanted to prune or care for it in case they did something wrong, or perhaps it was left in the futile belief that it would look after itself, as it had once done in the wild. But Rodrigues is no longer what it used to be—a cage with invasive weeds, thirteen feet away from a main road, is no longer the wild.

The roof and sides of the cage were smothered in a thatch of weedy climbers that blacked out the light. Inside, it was even worse. Just a few inches away from the trunk, a thuggish invasive species, the rose apple (*Syzygium jambos*), with a stem a foot thick, was bullying the café marron into submission, intent on suffocating it to death by taking its space and resources. The café marron, battling with mealybugs and other pests, had been staked with an iron and held upright with a rusty hook. It was like witnessing a king being tortured in a dungeon.

The café marron needed saving, again. Recalling the specimen luxuriating in the sunshine against a wall in Mauritius—happy, healthy, full of flowers, and beginning to develop fruit—I realized that, at the very least, this tree deserved to look like that. I turned to the forestry guard and, trying my utmost to stay calm, said, "Please can I do some gardening?"

"Oh no, no, no, no," he replied, "this is an important site—this is the wild."

"Ha! Listen!" I said. "There is nothing natural about this place! Though things have indeed gone 'wild.'"

I turned to Alfred. "Don't ask me," he said. "I think you're right—but you can't do this without permission."

I asked him who was in charge, and he gave me the name of someone else at the Mauritian Wildlife Foundation.

"Okay," I said, "let's go to his office."

Then Alfred and the forestry guard really started panicking. "Are you sure?" they said. "Do you really need to tidy the area?"

I tried to talk some sense into them. "You have to trust me on this one. Remember how healthy the *Ramosmania* was a few years ago? When the rose apple was half the size and there were no weeds?"

They agreed it had looked wonderful.

"I tell you how we'll do this. You give me permission, and if something goes wrong you can say that I didn't ask. How's that? Come on, I have fifteen plants waiting for you in Mauritius. Before I leave the island there will be many seedlings growing. If you want this to remain the oldest specimen we have, you have to let me do something. Left alone, it will die in the dark and its blood will be on our hands."

"Okay," they said reluctantly, "but don't be too heavy-handed."

At this, Alfred and I launched ourselves into the cage and began working feverishly. We ripped out the *Syzigium*, slashed down the weedy climbers and tore them from the cage. In poured the sunshine. We gave water to the thirsty café marron, treated its pests and gently removed the iron hook and stake. Then, as we cleared the ground, to our astonishment we found another plant, right in the corner and covered in weeds. It was so close to the fence that branches must have been cut from it when the cage was built. It was *Badula balfouriana,* one of only five specimens left in the world. It had been there the entire time. Alfred knew it existed, but it was a surprise to me.

❧❧❧

In Rodrigues I planted 600 seeds, which grew into about forty seedlings. We needed to start planting them out in the wild. Yet again, caution prevailed. The authorities started to make excuses: "What if they die?" They were racked with fear. I insisted, and we planted two with the team from the Mauritian Wildlife Foundation nursery. We set them inside the cage with their great-grandfather, creating a small population so that you could look after three for the price of one. We also planted some at Grande Montagne Nature Reserve on the side of a mountain where the ground had been cleared of alien species.

There was a local debate about who should plant the saplings—the Mauritian Wildlife Foundation or forestry officials—so I gave saplings and seeds to both organizations. The forestry guys used techniques that they were familiar with, planting everything equidistantly, in rows, through the existing weeds. Now the trees are in lines like a timber plantation, and hardly anything else is growing among them other than masses of invasive weeds. The MWF took a different approach. They worked out that if you remove the weeds, then plant native tree saplings, it doesn't help because the number of seeds in the soil means that the weeds will soon be back. But if you weed the area, wait, and weed again, it reduces the numbers of seeds waiting to germinate. They planted the saplings with several other species, with hardly a square inch of ground uncovered. These plants would start competing with one another, the strongest would survive, and because they were so close there would very soon be a dense umbrella of foliage, stifling the weeds below.

Now, when you visit that planting more than a decade later, it looks spectacularly random, with all the vegetative layers of forest in place. Different plants dominate in different locations: some become prominent where it is wet, others where it is dry. The whole thing looks natural.

Even more pleasingly, the Forestry Department—who were used to planting exotic ornamentals like hibiscus, cordyline, heliconia, and strelitzia, by public demand—have changed tack and are starting to value their own native species, which is having a huge impact on the island and its plants.

The plants in quarantine were released from Mauritius a few

years later and they now have females in flower from the first batch of 600 seeds that were sown on my first visit. Some of the staff from the Mauritian Wildlife Foundation at Rodrigues have been over to Kew, where I have guided them through the cross-pollination process—from how to recognize the male and female flowers to when to harvest the fruit. They have everything they need to continue a conservation program on the island.

People often ask, "Did you ever meet that teacher, Raymond Ah-Kee, or his student, Hedley Manan? They should be local heroes." On my first visit I asked again and again to speak to Hedley. I wanted to say thank you, to tell him that he was the reason I could bring these plants back. But people made excuses: "We'll see, we'll see." I suspected something was wrong. Then someone told me plainly, "He had problems with drugs or alcohol. Hedley is dead."

He saved an endangered species from extinction, but, sadly, no one could save him.

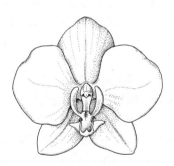

4

The Messiah in Mauritius

Black River Gorges National Park offers perfect post-card views of Mauritius: mountains on the horizon, pristine valleys covered in lush vegetation, waterfalls spilling white plumes. Here comes a luminous parakeet or white, streamer-tailed tropic bird, framed against the piercing blue sky. But it is a false Eden: nearly every single plant you can see, from those by your feet right out to the horizon, where the mountains disappear and the island meets the sea, are alien species. They are invaders.

The human history of Mauritius is also one of invaders. Even though it is so remote, it is one of the most cosmopolitan places I have ever visited. The dominant community is Hindu, sent from India by the British colonists to plant sugarcane. Before that the French introduced French Creole slaves from Africa and Madagascar. The Dutch, slaves from Java, and Chinese have lived there too. The dominant language is French Creole, but most residents are bilingual and speak English in addition to the language of their country of origin. In a small village, you may find a Christian church, a Muslim mosque, and a Hindu temple close together, or a Hindu guy having dinner next to a Muslim guy in a Chinese restaurant, all living in harmony.

Essentially it is an Anglo-French culture that was a Western colony by design. Even the botanic garden once had a French name. It was first Le Jardin du Roi at Pamplemousses, then Royal Botanic Gardens, Pamplemousses, and is now known as the Sir Seewoosagur Ramgoolam Botanical Garden. It's the oldest botanic garden in the southern hemisphere and well worth a visit, not least for its pond, which looks as big as twenty Olympic swimming pools and is filled with *Victoria amazonica,* the giant waterlily from South America.

Given this history, plant collecting there can be a political business. Wary of their biodiversity rights and the dangers of bio-piracy, the authorities in Mauritius are highly protective of their plants. The problem is they don't have the facilities to propagate all the species themselves, but they tend not to trust anyone else to do it either. So plant collectors and conservationists, not to mention the endangered plants themselves, are often caught in the middle of politics when the very existence of the plants is at stake.

It's not that the authorities don't have every reason to be concerned. The story of the critically endangered *Hyophorbe lagenicaulis* (bottle palm), for example, is a fascinating example of what can go wrong. This palm was once widespread on Mauritius but now exists naturally only on Round Island, some fifteen miles north of Mauritius, and there were once fewer than ten mature plants left in the wild. The seeds were in high demand: there were rumours of sheikhs in the Middle East who would pay a fortune for one. The minute somebody managed to get some seeds off the island, the plant was propagated, and it is now popular in front gardens in Miami. Mauritius did not get a penny, and yet they have to foot the cost for the conservation, propagation, and reintroduction of the species, which today is one of the iconic sights of the island.

Let's say that in the future one of the therapeutic benefits attributed to the café marron by traditional medicine is found to be effective. A major drugs company develops and patents a product based on the plant, making millions of dollars. The Republic of Mauritius and its islands would not receive a cent. It is not a surprise that countries like Mauritius, Indonesia, and Brazil—where thousands of useful plants are found in the rainforest and the shamans and indigenous peoples have a detailed knowledge of their uses—would want to stop this from happening. But the law created to help them is blocking conservationists from saving endangered plants.

The 1992 Convention of Biological Biodiversity, followed by the 2010 Nagoya Protocol—conventions relating to ownership of biodiversity and plant rights—inevitably made plant collection more slow and bureaucratic. The Mauritian government went a step further and challenged Kew by saying that the rights to any plant that came from Mauritius, whenever it was collected, were owned by Mauritius. It's no surprise there was a standoff, because Kew already had a large collection of Mauritian plants. The Nagoya Protocol covered only wild plants that were collected *after* the date of the agreement. If Kew bowed to Mauritian pressure, it would set a precedent for all countries to do this, despite it not being formally agreed. This disagreement halted our collecting and conservation programs in Mauritius for several years until it was eventually settled.

Time passing, and the repatriation of the café marron, eventually helped collaboration to resume. Plants were under threat of extinction, so I needed to be there. A species could have been around for millions of years, but when it is down to the last plant left and you can hear the sound of chainsaws getting closer, something has to be done.

※ ※

As part of my preparations for my first trip to Mauritius in March 2007, I did a lot of research, drawing on books and conservation websites to make a wish-list of places to visit. When I arrived, what I actually did was pal up with the people in the National Parks and Forestry Departments. They had the transport, knew where the plants were hidden, and could issue permits. Whatever the historical differences between Kew and the Mauritian authorities, they pulled together when I arrived. The Mauritian Wildlife Foundation and the Mauritius herbarium made me very welcome too.

The first thing the National Parks and Conservation Service officials said to me was: "So, what do you want to do?"

My main mission was to collect and propagate threatened plants, but they had to be able to carry on the work and cultivate the plants I had propagated—otherwise my efforts would be in vain.

Some of the items on my list of "must-see" plants could be propagated and left in Mauritius, while others would come back with me to Kew. These were often the last plants of their kind left in the wild. Each new plant I could propagate was a step closer to survival.

The conservation issues of the flora in Mauritius are numerous and desperate. We have to start somewhere, so let's look at the palms. There are seven species of palm on Mauritius, five of which are found nowhere else in the world. Palms were once everywhere on the island, but locals had a taste for their growth points, and "palm cabbage" become a popular lunch. Now almost every palm is under threat, with many only just saved by cultivation.

One of the first palms I went to see was *Hyophorbe vaughanii*. At one time there were only three left in the wild. Luckily, though, they produced a small quantity of seeds and a few nursery-raised saplings are now thriving in natural reserves. We visited the original wild ones at Florin Reserve, in the Black River Gorges National Park, which could only be accessed by squeezing the jeep along a narrow path cut through the encroaching guava thickets. To my surprise, the three specimens were growing so close together that the tops of each palm frond were touching the next plant; they were like three musketeers fending off extinction. No seeds were available, but they were in flower, their long, hanging inflorescences wearing blooms of white and brown. When we visited the population of saplings that had been planted in the wild, there were three seeds, so I popped them into my pocket and returned with them to Kew, where they germinated. One is now on display in the Palm House, and the others are in the nursery, where I am trying to wait patiently for them to reach maturity and produce more seeds.

Another palm that is endangered in the wild is *Hyophorbe verschaffeltii*—there are only thirty-six wild palms left on Rodrigues, and only four occur in an area with some protection. Nowadays it is widely cultivated around the world and on the island, but all the specimens come from those last few plants. This trip was an ideal opportunity to go back to the originals and collect seed from the oldest plants, so the DNA would be true to type—the best seed to store in a seed bank.

Next is *Latania verschaffeltii*—there are about 200 specimens left in the wild. In cultivation it is known as the yellow latan palm because the leaf stalks on young trees are yellow and the leaves are yellow-green. Despite the numbers, there is hardly any natural regeneration, so the Mauritian Wildlife Foundation and the Forestry Service are planting nursery-grown saplings in its native home, Rodrigues.

The wild population of *Acanthophoenix rubra*, the red or

Barbel palm, has striking red leaf stems at the top of its trunk with long, dark spines, which are shed as the leaf matures and the stems become browner. It has been reduced to 150 individuals on Mauritius Island as its habitat has been destroyed to make room for sugarcane plantations, and it has a high value as a medicinal and edible palm. Again, it is now only really grown as an ornamental plant in gardens, and on plantations for "palm hearts" (for more on which, see later).

The saddest tale is that of *Hyophorbe amaricaulis*. There is only one surviving palm in the world, in the Curepipe Botanic Gardens, Mauritius; it is not clear if it was planted there or left from the native vegetation. The species was described (that is, officially named and known to the scientific community) in the 1700s from specimens taken from Pieter Both, the second-highest mountain in Mauritius, where it seems to have been widespread. It is about forty feet high, with a relatively thin trunk of less than forty feet in diameter.

Of all the plants in Mauritius and Rodrigues, *Hyophorbe amaricaulis* is the one I am most concerned about. It is known as "the loneliest palm tree on the planet."

Everyone talked about Lonesome George, the last remaining Pinta Island tortoise in the Galapagos, who was alone for forty years before he died, because he was an animal. When the palm dies, what are we going to do? Will there be reports on television and in newspapers—will it get the same coverage? I very much doubt it. Whenever I think about it, I feel sick. I should be phoning the BBC. I should be knocking down the door of the Kew director's office, saying, "We need to fix it and we have to put the cash on the table." I should be doing more.

The problem is there are so many plants that need saving. However much this one palm means to me, all of my efforts and budget can't be spent on this plant alone. It would take at least four trips to Mauritius to save one plant when that time and money could be used to save fifty others. But that still doesn't make me feel any better.

A plant has never become extinct while I have been working on it. The day a last plant dies before my eyes, I am going to bitterly regret it.

I have visited the palm a few times on my trips, out of interest and sympathy. It only ever seems to have three to five fronds, and it is surrounded by scaffolding so scientists and horticulturalists have access to the flowers. There was great debate about this structure. Some believed that it would protect the palm from cyclones while others feared it could be decapitated by the frame during a cyclone. The main problem endangering its survival, though, is a flowering quirk: the male flowers open long before the female, making it impossible to pollinate. As it's the last specimen left, it can't produce seed without help. When I first went there it had a branch full of ripening fruits, so I was optimistic. Someone trained the botanists in Mauritius to collect the pollen and store it at the right temperature, so when the female flowers bloomed they could pollinate them and get them to set some seed.

Then came disaster: the island was hit by a cyclone, the branch snapped, and they lost them all. Other inflorescences (groups of flowers arranged together on a stem) set fruit later on, but none of the seeds germinated. A similar fate befell a later attempt. But before the seeds ripened, the staff at Curepipe sent some of them to the Micropropagation Unit at Kew, which managed to grow one plant. It grew up to ten inches tall, in a long sterile flask, but eventually gave up and died.

It was as though, after all its struggles, the plant had lost the will to live.

Some seeds had formed again when I arrived there for my second trip, in April 2010. I asked to have some. Since we had partially succeeded at Kew, I thought we should have another go; this might well be its last chance for survival. After complex negotiations with conservation organizations on the island, they finally agreed.

I explained exactly what I was going to do. My intention was to cut the fruit the day before I flew back to London. It had to be removed from the branch while leaving a bit of the stem attached, to prevent external bacteria from sneaking into the seed tissue and contaminating the embryo, making it unsuitable for the sterile culture we were going to attempt at Kew. The seed with stem attached would then be placed in a sterile bag, and I would rush it to one of the few "flow benches" (a device that filters all life forms from the air) on the island, which would allow me to sterilize the outside of the seed, before sealing it in another flask ready for the long-haul flight to London. As soon as I arrived at Kew, someone would be waiting in the Micropropagation Unit, with the flow bench switched on, ready to carefully extract the embryo.

The staff from the National Parks and Conservation Service, who were following advice from the staff at Curepipe Botanic Gardens, listened carefully and said, "No, leave it to us and we will cut it for you."

I had to give them a chance. I went out collecting for the day, giving strict instructions that it should be put into the fridge in the mess room to await my collection.

That evening, I returned. When I walked through the door, there was one of the garden laborers, chewing heartily and spitting husks onto a polythene bag.

They had promised me five seeds; there were only three in the bag.

"Where did you get those seeds you are eating?" I asked him.

"On the island, we like to eat palm seeds. I have never eaten this species before," he replied.

I wanted to throttle him. But I was so stunned, all I did was ask, "Did they at least taste good?"

"No, they weren't ripe," he replied abruptly.

Later, I discovered how a chain of errors had meant the seeds weren't looked after properly. The process had been riddled with misunderstandings. The garden laborer was probably unaware of the problems this species faced. He hadn't had lunch, had a taste for palm seeds, and, seeing them in the fridge, tried his luck.

It was embarrassing for all of us. There were only three seeds left. The sterile protocol was also a shambles.

When I calmed down and thought rationally, I realized that they were desperate to help and had tried their best. They were passionate about conserving the palm and just wanted a piece of the action too. A lesson learned. Despite their best efforts, the Micropropagation Unit was unable to do anything with the remaining seeds and they died.

One day, when we are successful, we will be able to laugh at this—I hope. But this palm could be just one cyclone away from extinction. If this plant dies, the species will be gone forever. While there is life there is hope, but time is running out.

❧❧❧

Once there were only two specimens of *Dictyosperma album* var. *aureum* left in the world. This elegant palm with golden fronds is found only on Rodrigues, and the two plants left in the wild were in Montagne Charlot, about a mile from Port Mathurin, the main town on the island. Now it is widely planted because people introduced seeds into cultivation, but seeds from plants of wild origin are a must for conservation.

When I arrived, I found that both specimens were on private property. The owner had made a flower bed underneath them, where she had planted busy Lizzies (the common name of

Impatiens walleriana) and patiently weeded out the seedlings of the palm to stop them from spreading. She was actually doing this when I arrived: it had to be one of the most ridiculous threats to a species of all time. The rest of the seeds were suffering death by mower, as they landed on the lawn of the garden. Nowadays the Mauritian Wildlife Foundation raises seedlings and distributes them around Rodrigues.

As if that wasn't bad enough, its close relative *Dictyosperma album* var. *conjugatum*—found only on Round Island—was down to a single palm, but they have now been propagated and, after a few decades, this species is now widely distributed in cultivation.

In fact, *Dictyosperma* palms are made of tough stuff. They are masters at surviving heavy winds; in their native areas they have evolved to withstand the destructive Indian Ocean cyclones. Just before the winds grow strong enough to topple any palm, *Dictyosperma* will lose all its fronds, leaving just the trunk standing, with the growing bud undamaged at the very tip, so that there is almost no surface for the winds to attack. For this reason they are now planted throughout the world, and it is one of the few palms that you can expect to survive Florida's worst hurricanes. A well-aimed machete strike, however, will take one down—no problem.

At one point in Mauritius there were maybe half a million *Dictyosperma album* palms. They were nearly all felled for "palm hearts"—a traditional dish. Now there are fields of them in cultivation, and they are to be harvested only when they are about seven years old. This could have been done before this palm became a threatened species, but it was easier just to cut them from the wild. Currently, when you travel around Mauritius, it is a common sight to see commercial crops supplying the demand for culinary use.

What a sensible idea. More of those, please.

❖❖ ❖❖

Chassalia boryana, or Bory's coral tree, is as beautiful as its name: waxy-white, with star-shaped flowers at the tip of each stem; the inflorescence is like a candelabra. Individual flowers open for just one day, but there can be six, sometimes ten, flowers open at a time. From the top it looks like a snowflake, constellation, or supernova. Its elegance changes every day.

I only came across it by chance. At the end of my first visit to Mauritius, just as I was about to jump in the car and leave for the airport, someone from the MWF gave me a paperback of the recently published *Guide to the Plants in Mauritius.* This is more like a visual guide, with pictures of the most common plants of the island and a few rare ones.[2] As I flicked through the pages, I came across a picture of a plant I had never seen before.

Chassalia boryana was named by Swiss botanist Augustin Pyrame de Candolle in 1830 for French botanist Jean Baptiste Bory de Saint-Vincent, who traveled around Mauritius and Réunion at the end of the eighteenth century. Despite being named by and for two famous historical figures in natural science, with names nearly as colorful as the species they worked with, the plant was not spared the calamities of Mauritius. For a long time it was believed to be extinct.

Though it is called the coral tree, it is in fact a shrub. It grows on a single stem, like a standard rose, and flowers at a height of about four feet, well above the mouths of the now extinct tortoises. When it flowers for the first time and that flowering tip dies, it divides into two or three shorter branches, repeating the process time after time and eventually forming a perfect lollipop shape. In the wild, the plants have lichens on their stems that look like watercolor marks, and this makes the shrub even more attractive.

The plant wasn't in the species descriptions, but I found it among a series of thumbnail images that adorned the margins of the book. I didn't recognize the plant at all, so I had to ask what it was.

"It's Bory's coral tree," said the person from the MWF. "It was thought to be extinct until the 1960s, when one plant was found, so we've known about it for a few years now."

"Where is it? Why didn't you tell me?"

"Too many plants, too little time. Also, we know very little about it at the moment," came the reply.

As the plane took off over the island, I had the book in my hand and was looking at the picture, then at the landscape below me, then at the picture. I could see the area where it was growing from the plane; I had fallen in love with this incredibly beautiful plant.

Like the café marron, this species is a member of the coffee family, so having a single clone did not help the prospects for reproduction. It was like discovering another *Ramosmania rodriguesi* from the window of a plane. If there had been plenty, they could have sent me some seeds in the post, but when there is only one, and I was in love, I could not entrust it to others. It needed my close, devoted, and undivided attention.

And it was the perfect excuse to come back.

As soon as I landed the second time, I got straight to the point:

"Good morning, nice to see you again. Where is *Chassalia boryana*?"

There are about six different species of coral tree in the Mascarenes. Each one is beautiful, and, as is often the case, the most beautiful of them all was also the most endangered. The fact that it had been rediscovered was extraordinary.

After the first *Chassalia boryana* appeared, extensive searches revealed up to fifteen more. When I went out to see the newly discovered population with a couple of representatives from the National Parks and Conservation Service, I said to them, "Look, I really need to take more than one cutting from more than one plant; like many others, it may have male and female flowers on different plants."

They were not very keen, and I could see their point—it's difficult to take a cutting without spoiling the shape of the plant. So I looked carefully and eventually found some plants where a cutting could be taken without interrupting the symmetrical shape. Other young plants were just single stems, yet to branch.

One day you are looking down at a beautiful forest from an airplane window, wondering where that last plant is hiding, the next day you are inside the forest itself, being serenaded by a Mauritius flycatcher sitting on the only sun-bathed branch in a shady thicket. While I was pondering which cutting to take, I heard an angry voice.

"Hey, what are you doing here? This is private property. Get out, get out!"

The National Parks and Conservation Service guards seemed to be under the impression that they had the power to enter any property to search for species and could collect whatever it liked, but this chap looked in the mood to challenge them. One of the representatives from the National Parks Service called the main office and told them to speak to the landowner, on speakerphone. They started shouting at each other, with me in the crossfire and potentially in hot water with Kew, because we are not allowed to collect without permission of the landowner.

As the argument raged in French Creole, the two people from the National Parks Service team urged me to tag the material. When I hesitated, they shouted even louder: "Do it, Carlos; do it!"

Luckily I realized that the two best plants were just outside the landowner's boundary, on public land, and said I would prefer it if we could take them from there. Conflict solved. I tagged them and we left.

By bringing the plant into cultivation, I discovered something else. I thought the flowers were going to be purple, as in the picture. But the first cutting had pink flowers. I also noticed a style (the female part of the flower that transfers the pollen's genetic load to the ovary), which often looks like a thick hair, projecting from the first flower. On the second day, after the first flowers wilted, I opened one of them up and found some stamens (the male part of the flower, made up of the filament and anthers, which produce pollen), but they were black and pollen-free.

"Hmmm," I thought. "I think I have a female here. Like *Ramosmania rodriguesi*, there may be plants of different genders."

The second plant to flower came up blue and male. I had a pink lady and a blue boy; if I managed to combine them, the beauty and attributes of the offspring would be at the top of the charts. I managed to cross-pollinate the plants and only one seemed to set seed—the one that was pink, which, in turn, had shorter anthers that never set pollen. However, the lack of pollen in one plant, and the shorter female parts in the one that shed abundant pollen, seemed to indicate that the plant was indeed dioecious—that is, there are male flowers on some plants and female flowers on others.

I suspected that because of the candelabra shape of the plant, the colors of the flowers, and the fact that they were upward-facing, with a tube leading to the nectar, the coral tree was pollinated by butterflies. The upward-facing black berries suggested the fruits and seeds were eaten and dispersed by birds. (If they were eaten by terrestrial animals, the fruit would fall and face downward.) There was also a color contrast—the coral candelabra was white and the berries black—which highlighted the berries.

When I sowed some of the seeds, however, which were like small coffee beans, up popped some white, mauve, pink, and purple forms. In the end I found the colors could be mixed. So much for my theory of "pink lady" and "blue boy."

The seeds of this species are "recalcitrant," meaning they do not survive drying and freezing, so they can't be stored in a seed bank. The only thing we can do now if we want to make sure it doesn't disappear is to cultivate the plant. I passed on the information about pollination and propagation to the locals involved in its conservation. Like the café marron, it has huge potential as a garden plant for the tropics, which is another way of preserving it. And anyone can now see the plant in London, in its glory, at the Palm House at Kew.

⇛⇚

Imagine a plant that is part liana and part shrub, with mangrove-like roots—how strange would that be? Let me introduce you to the critically endangered *Roussea simplex,* from the wet, high-altitude forest on Mauritius. At the moment it is in its own family, so if we lose it, the genus, species, and family will instantly become extinct.

In 1937, Reginald Edward Vaughan, a British botanist who lived in Mauritius, and scientist Paul Octave Wiehe wrote in the *Journal of Ecology* that this species was widespread. "In other places an extremely thick canopy of woody lianes (*Roussea simplex*) develops about 4–6 meters above ground level, causing such dense shade that both terrestrial and epiphytic plants are practically excluded," they said.[3]

Picture that. Nowadays it has almost vanished.

After an exhaustive search of the island in 2003 and 2004, fewer than ninety were found. One group in the north of the island, at Le Pouce and other small localities, contained about eighty-five plants; the other, a long distance away in the south, in Pétrin, an area of heathland in the Black River Gorges National Park, had only three. In 2007 I visited these three plants. On my second trip to the island, though, just two were left—while one was very healthy, the other was being overgrown by a massive screw pine (*Pandanus*), threatening its survival.

This dramatic decline is attributed to deforestation; the introduction of animals like rats, pigs, and monkeys, which grub up or eat the seedlings; and invasive plants competing for space—not to mention one other, odd factor, which I will come to later.

One of the healthy plants I saw was just teeming with life. You could spend a whole afternoon looking at this plant; it was loaded with flowers and fruits both times I visited. The flowers are complex, the fruits unusual, and the ecology incredible. Nearly every branch was loaded with orchids, lichens, and mosses, giving me an idea of just how much it was relied on by surrounding species. One of the orchids making itself at home on the *Roussea* was a lovely blooming specimen of the Mascarene endemic orchid *Cryptopus elatus*—a real stunner. It has a pure-white flower, which looks like one of those intricate paper snowflakes that children cut out with scissors at school.

Roussea simplex is a unique plant in many ways. Besides being the sole member of the subfamily (*Rousseoideae*), it is the only plant in the world that relies on the same animal for both pollination and seed dispersal—the rare Mauritian blue-tailed day gecko (*Phelsuma cepediana*), which lives mostly in the prickly-leaved screw pines, drinking the water that pools at the base of the leaves and eating the many insects that also live there, while protected from predators. I imagine it has a pretty good life.

But as the same animal is relied upon to pollinate and disperse the seeds, this makes the plant extremely vulnerable: if the pollinator and seed-disperser goes, the plant goes too. In fact, no other plant does this, for just this reason. A single extinction of the helper, and all of a sudden there are two problems to solve.

The plant itself has long stems but can be bushy too, and bears bright orange hanging flowers with thick, waxy petals that produce lots of yellow nectar for the gecko to drink in return for pollination. When the fruit develops, it is like the teat on a baby's bottle, and secretes a jamlike substance from its tip, full of seeds. The gecko licks the sweet gel, then disperses the seeds in its poo. The gecko never ventures farther than 160 feet from home, so for the plant to be pollinated and disperse its seeds it has to be close to a screw pine—a kind of lavatorial symbiosis. This intricate ecological association seems to be necessary for *Roussea simplex* to have a fulfilled and happy life.

The first time I saw the plant, I took cuttings, including one piece of stem where a root had appeared naturally due to the damp conditions. "This is going to be easy," I thought. But my optimism rapidly disappeared. Some of the cuttings stayed alive for about a year but never produced roots, so eventually died. The piece of stem with a root (known in horticulture as a naturally layered plant) was little better. Although it had a single large root, it did nothing, then died. It was very strange.

I tried again in 2010—this time from seed that I had brought back with me from my second trip, seed that the staff from the National Parks and Conservation Service told me would be difficult to germinate. I had also emailed my friend Dr. Viswambharan Sarasan in the Micropropagation Unit at Kew while I was in Mauritius, and he suggested something other than the traditional method of sowing seed when ripe: "Do exactly what you do with the orchids—collect unripe fruits, take them to the flow bench, sterilize them, and put them in flasks." I did this, and the unripe fruits arrived back at Kew with me, less than thirty-six hours after collection.

I decided to try to sow the seeds from ripe fruit in conventional ways, since there were plenty of them. The challenge, though, was how best to clean them. The sticky jelly had to be removed to prevent the seeds from rotting. In the wild, this jelly would be digested by the blue-tailed day gecko's gut. What I needed was a bit of lizard mimicry.

I tried spreading them out on paper to dry, but that didn't work very well, as the jelly formed a thick crust rather than evaporating, so I went for a cup of coffee and thought again. The second time I squeezed a massive ball of jelly out of the fruit so I could get plenty of seeds, and put them in water. By adding water, then decanting, over and over again, a bit like panning for gold, I managed to wash away the jelly successfully and was then able to sow my "gold dust" seeds.

Only one germinated, but with this washing technique I became the first person in the world to germinate a seed from this plant in cultivation.

A year later it was not doing much—just like the cuttings, it was going yellow and chlorotic. I decided it was probably lack of nutrients and fertilized it a tiny little bit. As a result it died. That's a drastic way of showing your disapproval. At least I learned something from its demise, though.

The plants that Sarasan had managed to grow in the Micropropagation Unit were now doing well, though. No one knew if the plant I had collected the unripe fruits from had been cross-pollinated by another, so there was a chance that most of the seeds would be sterile. Germination was poor, but we still managed to keep alive those growing in sterile conditions, and multiplied them by division too. I also contacted botanist Claudia Baider, who works at the Mauritius herbarium, to ask if she would collect me some seeds from the other end of the island, where there was a larger population. Together we could gather a much greater volume of seed and get some genetic diversity in the offspring.

With the original plants from Micropropagation and those that survived from a second batch I germinated, we were at last gaining some momentum, though it had taken four or five years. But then this second batch died during a hot summer and the number in cultivation crashed once again.

Maybe the species was sensitive to high temperatures, I wondered. After all, the small population in the south was in a cooler part of the island. But then they *are* from Mauritius, and the article said that they used to be all over the island—they should be able to take the heat.

It was only then that I had the thought: "Perhaps it is because the seed was taken from the population of three, at a high point on the island. That population may be more sensitive to heat."

I needed to find out about the conditions at Le Pouce, where most of the rest of the plants were growing, and Claudia Baider again came to the rescue. She said this population was found in a southeast-facing position—the equivalent of northwest-facing in the northern hemisphere—and that Le Pouce was rather cool, so high temperatures, even in an English summer, could indeed be too much for the plants.

I experimented. The following summer of 2013 I put some tiny plants inside an air-conditioned cabinet and they all survived. Far from liking it hot, as expected, they just like to be cool.

Perhaps this plant has been alerting us to the fact that global warming has been happening for much longer than we thought;

the plant and its ecosystem have been nearly wiped out by a warming world. Claudia also reckoned it germinates only at the base of tree ferns in its native habitat because it is moist there and free of competing weeds, and the plant has adventurous roots that end up growing all over the place. Even then it would still need the gecko to deposit the seeds.

There is a further twist. Not all the damage on Mauritius has been done by large grazing animals like sheep and goats. Settlers accidentally introduced a tiny ant, first described in Indonesia in 1861—*Technomyrmex albipes*, found in the Indo-Australian region, from India to eastern Australia and throughout the Pacific—which forages on nectar and fruit pulp. The ant has discovered that the hollow flowers of *Roussea simplex* last for a few days, so it puts mealybugs inside the flowers, seals them in with clay, then farms them for "honeydew" in captivity. When the gecko comes to pollinate the flowers, the ants attack the lizard to drive it away and the critically endangered *Roussea simplex* isn't pollinated. If, after a constant ant attack, the gecko stops visiting a plant, it will stop reproducing through lack of seed.

When you think of threats to endangered plants on Mauritius, you would never think of an ant. Human-introduced eco-bombs strike in the most unexpected places.

Luckily, the plants are finally doing well in cultivation, surviving with careful monitoring inside the protected fenced reserves of Mauritius and the glass nurseries of Kew. Hopefully one day they will break free.

One final thought: *Roussea simplex* was named after the intellectual Jean-Jacques Rousseau, an eighteenth-century Francophone-Genevan philosopher, writer, and composer. I am an admirer of his political philosophy, and he was a fan of natural science too. While reading Rousseau, I came across this quote:

The first man who, having fenced in a piece of land, said "This is mine," and found people naïve enough to believe him, that man was the true founder of civil society. From how many crimes, wars, and murders, from how many horrors and misfortunes might not any one have saved mankind, by pulling up the stakes, or filling up the ditch, and crying to his fellows: Beware of listening to this impostor; you are undone if you once forget that the fruits of the earth belong to us all, and the earth itself to nobody.[4]

Fences were once used to protect private property and livestock from wild nature, but nowadays fences protect wildlife from us. Funny, that. I wonder what Rousseau would have thought about the plight of his plant.

5

Talking Tortoises

In 1708, François Leguat described the most efficient way to move around Rodrigues Island—by jumping from shell to shell on the native giant tortoises.

Tortoises are one of the oldest living animals we know. Fossil records indicate giant tortoises once inhabited every continent and many islands, with the exception of Australia and Antarctica. They are remarkable beings, playing a similar role to elephants in the African savannah now: dispersing seeds, fertilizing the ground with their manure, and even creating walking tracks for other creatures to use.

Sadly, the benefits were all one-sided. For these and other native fauna, the arrival of Europeans was like a nuclear bomb.

The islands became staging posts for long voyages, where sailors would take a break and stock up with supplies. On Mauritius, travelers released cows, pigs, goats, and sheep—and with no predators, their populations exploded. Then the British, French, and Dutch arrived on ships riddled with rats, adding to the population that was already established, which had swum ashore from Portuguese shipwrecks. Settlers introduced invasive weeds and felled the forests for grazing (just like in Brazil today). Wherever Europeans went, they seemed only to colonize, exploit, and destroy.

Wildlife in many countries had already felt the impact of humans in an entirely different way. Polynesians had landed in New Zealand, Aborigines in Australia, and First Nation peoples were in North America, yet compared to our colonizing civilizations, their impact on the environment was at a slower pace and on a smaller scale. They may have arrived with crops and chickens, but not with enough plants or animals to create such a negative impact. Their populations remained small and they lived more in harmony with the native wildlife. True, they caused some extinctions through hunting, like the moa in New Zealand—a flightless bird similar to an ostrich but twelve feet tall and weighing 500 pounds. But those sad instances were comparatively few and far between.

European settlers destroyed the balance between nature and man. Indiscriminate grazing and logging, the introduction of invasive weeds, and ground clearance for agriculture destroyed the native flora. Some 50 percent of Mauritius was covered by sugarcane, and it is estimated that 95 percent of its native forest has been destroyed and only 1.6 percent of native vegetation is in relatively good shape. There was wave after wave of extinctions, including such biological wonders as the giant skink (*Didosaurus mauritianus*), a species of lizard that became extinct about 1600, leaving only an incomplete skeleton as proof of its existence. A mostly ground-dwelling black parrot that was about a foot and a half long is also extinct. It had a big beak, which helped it crack the large seeds of the blue latan palm, *Latania loddigesii*, a palm species that was totally exterminated on mainland Mauritius, surviving only on offshore islets. Then there was the infamous dodo. All of these and more became extinct on the island because they could not protect themselves against human activities and the invasive species that were introduced.

For the giant tortoises that Leguat rode, it was a disaster. Rats, pigs, and macaque monkeys contributed to the loss of eggs and juveniles. But things got worse—sailors discovered that the legs of tortoises could be removed individually and they would still stay alive. You could even make soup with the head, and once it had been removed the heart would still beat for a few hours more. Tortoises would be stacked upside down in the bottom of a galleon, providing fresh meat through the long months of a voyage. They were also invaluable water sources, as they stored it in a special bladder; even after the tortoise was dead, the water was perfectly drinkable. HMS *Beagle*, the ship that took Darwin on his round-the-world trip and led to the biggest biological revolution ever, was stocked with thirty giant tortoises, but none made it to the UK: they were all eaten before they got there.

There were so many factors against them that extinction was unavoidable.

Of course, extinctions are usually pretty bad news for the animals involved, but they also pose an unexpected threat to flora. Some fruits were dispersed by tortoises that are now extinct. Those tortoises also grazed some species more than others, which shaped the balance of the vegetation. You may find twenty plants of a critically endangered species that produce thousands of seeds, but they don't go anywhere because the species that dispersed them is extinct. There may be a pollinator for a plant, but does it still exist in the isolated area where the plant is now growing, or has the surrounding habitat been lost? On islands like this, where habitats have been degraded, such ecology is like a fragmented puzzle.

Yet it can also change quickly for the better. If you reclaim land from invasive species and also remove the introduced mammals, which decimate areas with heavy grazing, the vegetation changes and native species reappear. The impact can be seismic.

When they did this on Round Island, in a project sponsored

by the Durrell Wildlife Conservation Trust, the keel-scaled boa (*Casarea dussumieri*)—the only snake of its genus and one of the rarest snakes in the world—increased from a population of about 250 to more than 1,000. This turnaround was achieved because the population of its prey—small reptiles—increased. It is the only vertebrate that can move its upper and lower jaw, because its main prey, Telfair's skink, a type of lizard endemic to the island, inflates itself like a ball when it is attacked. Though it was too late to save the Round Island burrowing boa (*Bolyeria multocarinata*) from extinction, it is clear that such actions can have a dramatic effect on the remaining surviving biodiversity.

The Île aux Aigrettes is a small, flat island of coral limestone, a half-mile off the southeast coast of Mauritius, which is now a nature reserve and research station. They eradicated rats and reintroduced Aldabra giant tortoises (*Aldabrachelys gigantea*), which are closely related to the original native tortoise species, as part of a policy ("taxon or analogue replacement") being undertaken by the Mauritian Wildlife Trust. The basic idea is that when an ecosystem collapses because an important element is missing, you recruit a different species to do the job. This tortoise has been introduced to Round Island too, and the results have been phenomenal. Here they have discovered that the tortoises prefer to feed on introduced weeds and most of the seeds don't survive the passage through their gut. They eat the native species only as a last resort, so these have time to regenerate and establish themselves rather than becoming tortoise food. What the tortoises are doing is effectively hoovering up the weeds, fertilizing the native plants with their poo, and dispersing their seeds. The native species are returning; it is creating a micro-Mauritius out of Mauritius.

Now 125 tortoises have been released on Round Island, a few miles from Mauritius, including one species native to Madagascar, the radiated tortoise.

Many tortoises are suspected to live to more than 200 years of age, but this is difficult to track since they tend to outlive their human observers. One, called Adwaita, was reported to have been one of four tortoises brought by British seamen from the Seychelles as gifts to Robert Clive of the British East India Company in the eighteenth century, and was taken to the Calcutta Zoo in 1875. At Adwaita's death in March 2006 he was thought to be 255 years old. Today, Jonathan, a Seychelles giant tortoise (*Aldabrachelys gigantea hololissa*), is thought to be the oldest living giant tortoise at the age of 184, while Esmeralda is the second-oldest at 170. Esmeralda is an Aldabra giant tortoise; some of her relatives are helping to restore Mauritius flora back to its former glory.

I had firsthand experience of giant tortoises when we landed on the Île aux Aigrettes. My first encounter was catching a couple of them sharing an intimate moment. I noticed them because of the noise they were making; let's just say it was ungainly. Later I was quietly taking photographs when I heard a hissing noise and then thumps, getting louder and faster in turn. It was a giant tortoise trundling toward me, about four feet long, like a tank. He looked sturdy, as if a garden tortoise had been crossed with a black rhinoceros. I gently jabbed him with my camera tripod but he would not give up. I thought I must be in his path, so I moved about twenty yards away, but still he came toward me.

Every time I moved, he changed direction. I went left, he went left; I went right, he went right. Everywhere I went on the island, this tortoise followed and tried to take me on. With temperatures of about 95°F, I was starting to get sweaty and fed up. Finally, out of sheer frustration and to make myself lighter, I took off my backpack and chucked it on the ground. Within seconds he had started attacking it.

A guy from the reserve appeared and started chuckling: this

tortoise was the biggest male on the island and was blind in one eye. As the dominant male he must have decided that I, plus backpack, was a rival. I walked around the island with the guy from the reserve for three or four hours, and when we came back the tortoise was still going at my backpack, trying to turn it over as though it was an enemy.

These not-so-gentle giants are also known to attempt perilous acrobatic feats, rising up on their hind legs to reach low branches—a high-risk job, as they can tip onto their backs and die if they're unable to turn over. This unusual behavior led Mexican biologist José Antonio de Alzate y Ramírez to refer to the Aldabra as the ninjas of the tortoise world.

They may also travel more often than you imagine. Ever wondered how a giant tortoise colonizes an island? By floating. In December 2004 an Aldabra giant tortoise was washed ashore on the coast of East Africa, probably having been carried off the Aldabra atoll, 450 miles away.

❖❱❰❖

Mauritius is awash with endangered species. Among them are about eighty-nine species of native orchids.

One of the most mind-blowing is *Angraecum cadetii,* which is critically endangered on Mauritus, though not threatened on Réunion. There are only about a dozen plants left in the wild; they are found at about a half-mile above sea level in a place that is like a heathland in humid forest, where they are watered by the fog and the rain. Because the ground is so poor, their host plants are shrubs, about six and a half feet tall. The orchids are generally found low down on the shrubs, about two to five feet from the ground, where they receive a steady supply of water and nutrients as the rain trickles down branches.

I found a beautiful specimen in flower at chest height in

a small native shrub: the waxy greenish-white blooms of the orchid were offset by its glossy, dark leaves, arranged like a Spanish fan. I told the team I was with how Darwin had predicted that *Angraecum sesquipedale* from Madagascar was pollinated by a moth with a long tongue, but no one believed him until the Malagasy subspecies of the African hawk moth was discovered a few decades after his death.

Darwin was impressed by this plant. On January 25, 1862, he wrote to Joseph Dalton Hooker (who later became director of the Royal Botanic Gardens, Kew): "I have just received a box full [of orchids] from Mr Bateman with the astounding *Angraecum sesquipedale* with a nectary a foot long—Good Heavens what insect can suck it."

When we found the *Angraecum cadetii* in flower, one of the guards from the national park turned to me and said, "Okay, come on, clever boy, how is this one pollinated?"

The flowers are dumpy, wide, and very close together, with thick petals and greenish sepals, but there is no nectar—what geckos usually want as a reward.

After pondering this for a while, I finally admitted: "Okay, I haven't a clue. But whatever it is that is pollinating the orchid, it visits at night and the packet of pollen in the flower is housed in the part of the flower that looks like the shape of a head—so looking at the size of it, this must be a guy with a very big head."

Something was obviously pollinating them, because they were setting fruits, and even though the orchid was related to Darwin's *Angraecum sesquipedale*, it certainly wasn't a moth: *Angraecum cadetii* doesn't have the sort of spur that a moth needs in order to suck nectar through its long proboscis.

Later I discovered that a colleague from Kew had already visited Réunion to research the same puzzle. They had gone out into the wild and watched the flowers all day for potential pollinators. Nothing happened. The next morning, when they returned, much to their surprise, one of the flowers had been pollinated.

They set up a night-vision camera and waited.

Finally, after several nights, the remote camera was triggered. When they pressed "playback" they watched in awe as a cricket crawled onto the flower, poked its head into the right spot, and drank the sweet nectar before leaving with packets of pollen attached to its head. It then moved on to visit other flowers on the orchid by climbing up the leaves, then jumped to neighboring plants.

As I had expected, the size of its head perfectly matched the size of the opening between the orchid lip and the "pollinia" (the name for the lumped mass of pollen that is found in orchids). Most crickets find a place to hide in the dark during the day, usually somewhere different every time, but this one would go back to its nest so it could find the orchids again. The cricket, *Glomeremus orchidophilus,* or the Mascarene raspy cricket, was a new species to science too. Crickets are better known for eating flowers than for pollinating them, yet this one didn't damage the orchid at all.

Angraecum cadetii is the only known cricket-pollinated plant in the world. If the cricket becomes extinct, the plant will lose its pollinator—just like the blue-tailed day gecko and the *Roussea simplex*, it would be a double-whammy of extinction.

The first year I went to the island there were a few specimens of this lovely orchid growing in pots in the nursery. The guys working there asked me: "When you put this back into the wild, how will you attach it to a branch or tree trunk?"

"Ladies' tights. We need to go to a shop and buy some tights," I replied, as if it were the most natural thing in the world. You could tell by their reaction that they thought I was mad—or planning a robbery.

I explained: "We take each leg and cut it into rings, like cala-mari, then snip each one to make a piece of string. You can then tie as many as you like together and use them to attach orchids to branches without cutting through their roots or the bark of the tree it is attached to. They need to be tied firmly, of course, but there is plenty of room to expand, and the orchids eventu-ally attach themselves to the bark."

I soon found myself alone in a lingerie shop in Mauritius, asking an assistant for ladies' tights, 30 denier, preferably brown (only the best for an orchid).

6

River Deep, Mountain High

An iconic Mauritian sight is Le Morne Brabant, a basalt outcrop 1,800 miles tall, its cliffs plunging into the sea. It sits at the end of a peninsula on the extreme southwest coast of the island, and is a strange yet haunting place, with the presence of Sugarloaf Mountain in Rio de Janeiro, Brazil, or Uluru (Ayers Rock) in Australia. The summit, which covers more than twenty-nine acres, is thought to have been a hideout for runaway slaves, who at least had spectacular views over the turquoise sea.

Le Morne Brabant is home to many endemic species, including the gecko-pollinated *Trochetia boutoniana,* the only known population of the national flower of Mauritius, with its bright red, bell-shaped flowers. It grows on the cliffs at the top of the mountain, with the blue sea beyond, and a sighting of it against this backdrop is particularly magical. The mountain is also home to *Hibiscus fragilis,* which we have at Kew. Unlike most hibiscus, which are trees or shrubs, in this location this variety is a flat, floppy thing, up to thirteen feet wide and no more than three feet high, with bright red, white-veined flowers. I have tried taking cuttings from more upright shoots, but all it ever wants to do is grow horizontally, like a pancake, even in cultivation. I always wondered why, until I climbed that mountain.

Each time I visited I told people I would like to climb it, but they were never keen and tried to warn me off. "The paths are narrow and slippery, especially after the rain," they'd say. "It's far too dangerous. There's no point in your going anyway, as you can't collect without a permit." That only made me want it more. I would gaze at the mountain from the beach at the end of every day, and it would call out to me.

I knew that conditions up there were harsh—hikers started early in the morning to avoid the searing heat of the day—but from a botanical point of view it was fascinating. I was desperate to see the plants it held in the wild.

One Friday I met a local landscape designer and orchid enthusiast, François, who had a good collection of tropical plants and tried to include native varieties in his work. When he found out I was "the man from Kew," he offered to take me up the mountain. Ignoring the warnings, and unable to resist the temptation any longer, I agreed to meet him the following morning.

Just before sunrise the next day, we drove along a road that wound around the base of the mountain, then left the jeep and began our walk along a track that zigzagged ever upward, among a thick tangle of invasive growth and local endemic plants. Every so often I had to shuffle along the narrow path like a tiny ant, my back pressed against the rock face, arms outstretched, and with only a fraying rope between me and a drop into oblivion. It was well worth the vertigo.

Along the way I found interesting plants, including the critically endangered *Hibiscus fragilis*. And, would you believe it, they were all pressed horizontally or vertically flat, as though they were stuck to the rock, just like our clone at Kew. Now I could understand why. In such an exposed position they needed to protect themselves against the harsh winds and cyclones that batter the island. When I mentioned this to the staff at the Mauritius herbarium, though, I was told that specimens from Corps de Garde, a half-mile-high mountain some distance away from Le Morne, grow in an erect manner. It seems that these growing habits are genetic: the seedlings will take the same form as their parents, whether there is wind or not. At Kew, the plants are from Le Morne and grow like those in Le Morne, while Claudia Baider from the Mauritius herbarium has observed that seedlings from erect forms always grow in an erect way. Same species, different tricks.

From a distance I had assumed there must be seabird colonies because I could see what looked like thousands of splashes of guano highlighted against the dark rocks. It turned out to be *Helichrysum mauritianum*, which forms ground-hugging cushions of a bluish-white color and is made of tightly packed rosettes of leaves that look like origami snowflakes, about two inches wide. Nearly all the plants grow in almost vertical, black, igneous cliffs, with the sea hundreds of feet below. Hiding up there too was the critically endangered *Distephanus populifolius*, a member of the daisy family, with clusters of bright yellow, sweetly smelling flowers, like a giant groundsel, and leaves covered with white hairs that reflected the sunlight and highlighted the veins. There are few populations left on Mauritius, but happily it is flourishing in cultivation and has been growing with us in the Temperate House at Kew since 1985.

At the very top of Le Morne is one of the last living specimens of *Badula crassa*. The genus *Badula* is in the plant family Primulaceae, making them relatives of the herbaceous plants primrose, cowslip, and cyclamen, so it was strange to find that *Badula* are woody plants, like shrubs or small trees. Only ten specimens of *Badula crassa* have been seen since 1995, and most of them have now died. François asked me, "Why don't we go to where this plant lives and then you can take a cutting?" But it was impossible: collecting without a permit would upset the people at the National Parks and Conservation Service. In a way, it was a relief too, because to get to the very top of the mountain you have to cross a vast crevasse on a tattered rope bridge.

Several different species of *Badula* are under threat in Mauritius; I am doing what I can for them, but it is not easy. Only six adult *Badula ovalifolia* plants have been found in Mauritius; four have since died, and now there are only two left in the wild, only one of which sets seed. It is shaped like a Christmas tree, with large leathery leaves, and bears clusters of tiny white flowers that shed masses of pollen when they are shaken. I have taken cuttings from it, and we now have several plants at Kew, including one on display in the Palm House. But *Badula reticulata* is more problematic. There are twelve plants of its kind left in the wild, and every time I take a cutting, the mother plant dies. I have been stuck with a single cutting and the same problem for seven years.

Finally, I left the mountain, satisfied that I had at least managed to see nearly all the hidden endemic jewels on the legendary Montagne Le Morne, leaving just the *Badula crassa* for another visit—ideally in a helicopter.

⟫⟪

Cylindrocline lorencei is an unimposing small treelet, reaching six and a half feet in height. It has rosettes of leaves that are soft and furry, like donkey's ears, when they are young. When I say "it," I mean exactly that: we've only ever known a single plant, which was discovered in Plaine Champagne in the Black River Gorges National Park. Repeat after me: *a single plant.* We never saw its mother or knew its father: they had been destroyed long before. It grew without family, silent and alone.

In 1980, when the botanist Jean-Yves Lesouëf went to

Mauritius on holiday, he collected a few seeds from this last plant and took them back to France. Germination tests in the Conservatoire Botanique National de Brest (Brest Botanic Gardens) in Brittany, in 1994 and 1995, failed as the seeds did not seem to be viable. No one is sure why, but they didn't germinate—perhaps they were dead, or sterile, or the growing technique was wrong. There were not enough seeds left in the world to experiment further. By 1990, *Cylindrocline lorencei* was considered extinct in the wild. All that was left were dying seeds in a laboratory.

Then something miraculous happened.

There are two main parts to a seed: the endosperm, which is the seed's food supply until the roots and shoots are formed, and the embryo, with all the plant information, such as the DNA—there are quite a lot of cells in there.

Even though appearances can suggest otherwise, a seed is actually a living organism. It is now possible to run a staining test, which highlights living and dead cells in different colors. Brest Botanic Gardens tested these seeds and found that most of the cells were dead, so the seed could not function naturally. But the cells don't *all* shut down at the same time; the decline toward death is a gradual process and there were still some living cells.

Very carefully, they extracted the living cells from the last three seeds. When growing plant material in micropropagation you can take the embryo or cells and grow them in a solution containing all the nutrients that a plant needs—it is almost like a test-tube baby or growing stem cells. It starts out as an odd-looking cluster of cells, a bit like a wart, then after a time a plant gradually forms. After several years' growth, they managed to get the first plant out of the sterile flask where it was being cultured, and eventually ended up with three clones.

They sent some plants to Kew in 2001, where they have been growing. We managed to root one *Cylindrocline lorencei* from a cutting, but it wasn't easy. In fact, most of the cuttings we take generally die. The plant does not branch that often, so it would be a slow way of propagating the plant anyway; micropropagation—where plant tissue is used instead of cuttings from larger plants—was definitely the most effective technique we had.

In 2006, micropropagated plants in jars were repatriated to Mauritius, but unfortunately the plants failed to establish while being weaned on. So a year later I personally took some plants back to the island. It was an honor, though initially I wondered if they would make it—at two feet tall, the plants were too big for hand luggage. Do you know that airlines have no provision for protecting plants from death or damage in the chill of the cargo hold? It's fine to send your cat or dog in a plane, but there is no service that guarantees they won't freeze your plants.

I made so many phone calls to the airline and packaging companies to solve this—you would think I was asking the earth.

"Okay, just theoretically," I said to one airline, "what about if I worked in a zoo? Could I put a tiger in the hold?"

I discovered that in that case there *is* a service. There is even a way to fly a cow.

"So what do I do? Put my plants in a cow-sized box?"

"I am not sure, sir, but the system doesn't allow me to do this."

Eventually I found someone who could guarantee the safety of my plants. I was told to arrive at the airport the night before, where the plants were checked, scanned, and put on the plane.

On arrival in Mauritius, in March 2007, my twelve precious specimens of *Cylindrocline lorencei* went into quarantine in the Native Plant Propagation Center. Their natural habitat was in the coolest place on the island, and they would not like heat in the summer in the lowlands, so I asked if they could be quarantined up in the highlands—in conditions similar to where the first plants were found.

"No," they said. "We have no facilities there."

Plants don't understand bureaucracy. They can't wait, and by the time the propagation center sorted things out months later, only two plants were still alive. One survived for a while and was sent to the Pigeon Wood Gene Field Bank in Black River Gorges National Park, a kind of arboretumlike nursery planted with rare endemics, which was much higher up.

I was sad that the others died because of their heat sensitivity and the quarantine issues; ideally, they would have had a short quarantine in the lowlands, followed by close monitoring at higher altitude in the gene field bank. These were some of the only specimens in the world. The authorities, however, were forced to confront the reality that things would have to be done more quickly if they wanted their flora to survive.

I am often researching and trying to propagate so many plants at once that I feel like a magician spinning plates—each time one stops, I have to go back, see why it has faltered, and start it spinning again. Seven years after the last repatriation attempt, I remembered that when we took cuttings from *Cylindrocline lorencei* they nearly always rotted. This is likely to be because, like many other woody plants in the Asteraceae, or daisy family, *Cylindrocline lorencei* have hollow stems filled with spongy pith, so they rot easily. As I said previously, in micropropagation at Kew the plant propagated quite easily—it was almost like a growing version of "cut and paste"—but my colleagues in that unit were using tiny amounts of plant tissue in sterile flasks.

I wondered about trying to take cuttings of mini-stems instead: it was likely that these stems would not be hollow like the more mature stems, and would root easily from cuttings. So I tried an experiment. I pruned back the main stems hard and waited for the small dormant buds on the leafless stems to burst into growth. When the shoots were almost an inch long, I took tiny cuttings with a scalpel from the very tip of the shoots. It was a kind of micropropagation without sterile conditions. I also replaced the agar that we usually used with fine shredded coconut fiber and washed river sand. Guess what? They rooted easily in the mist unit after only two weeks. Now I can propagate them whenever I want, though they root particularly well in spring. Using this technique, I have as much success as when the plants were micropropagated, and they can now be grown on the island. We just need to establish a few stock plants in the highlands. Arrangements are being made to start restoration work in the original areas where this species was found, so that plants propagated at Kew and Brest botanic gardens can be reintroduced to the wild. Hopefully one day *Cylindrocline lorencei* will be home to stay.

❖

Sometimes, pronouncing the final extinction of a species can be tricky. There may still be a single plant left somewhere, unnoticed, just like the café marron, when after forty years the last plant was found. Thanks to the Internet and social media, news now reaches the most remote places, and people can see what they are looking for on a screen, increasing the chances of people locating rare plants.

However, there's precious little hope that another specimen of café marron will ever be found. Rodrigues is quite small. There have been some incredible cases of rediscovery—the chances increase when there are so many rare or extinct species on a single island—but no new café marron to date.

There is one genus that seems to specialize in ghostly

appearances—the genus *Dombeya*. You would normally see this member of the mallow family growing as an ornamental plant in the tropical and temperate zones, as a *Dombeya wallichii*—a large open shrub with heart-shaped leaves and dense clusters of bright pink flowers. But other members of the genus have not been quite so fortunate.

Dombeya rodriguesiana, endemic to Rodrigues, was known from a single surviving specimen. It was propagated by cuttings, which in this species are actually quite difficult to get to root successfully, and two small trees survive—one in the local native plant nursery at Solitude, run by the Mauritian Wildlife Foundation, and another one at Brest Botanic Gardens in Brittany.

When the wild plant died, all we had left were the cuttings. To complicate matters, the species seemed to be dioecious, with male flowers on some plants and female flowers on others. This last wild plant was a male.

This was the story when I visited Rodrigues for the first time. But by the time I landed again, a second plant, a different clone, had been found. It was in an area called Anse Quitor, in the back garden of a local farmer, who kept chickens around the tree. This farmer said it had been there when she moved to the property a few years before. It was about eight feet tall, five feet wide, and had a trunk of about four inches in diameter. Once it had been identified it was fenced in with chicken wire. I tried to propagate this newly found plant, but the remoteness of Rodrigues meant that my cuttings spent too long between being cut from the tree and planting, so they never rooted.

A close relative, *Dombeya mauritiana*, from Mauritius, offers a greater hope of success. The plant has similar leaves to a lime tree and makes low globular shrubs that are as tall as they are wide—usually six to ten feet. *Dombeya* generally have droopy, rather large flowers, ranging from white to deep pink, but *Dombeya mauritiana* is radically different. The flowers of this species start off small and white, in small flattened flower heads, and then turn brownish and stay like that for a while before wilting. The plant was once known from a single male specimen in the wild, growing in the lowlands, in a hot and sunny part of Mauritius. This specimen never set seed, but cuttings were taken and these have been growing for years at Kew and Brest botanic gardens—though, again, they have never set seed.

On my first trip to Mauritius, however, there were loads of seedlings in the Native Plant Propagation Center in Curepipe, run by the National Parks and Conservation Service.

"Hey, what's going on?" I said to them. "I thought this thing never set seed."

"Oh yeah," they said, "the cuttings we took years ago were planted in the arboretum and set seed."

I noticed that the seedlings had small hairs on the undersides of the leaf surface. Some of these hairs were golden, like those on the original plant, but some were white and others were reddish.

I once tried crossing a plant from Réunion called *Ruizia cordata*, a member of the same family as *Dombeya*, with *Dombeya mauritiana*. Seeds were produced and germinated with ease. *Ruizia* was originally known from a single clone, and it was male too. In a further twist, this male was later reported to have changed gender and set seed. Go figure. It seemed that we had a plant that changed sex according to the growing conditions.

There are twelve to fifteen species of *Dombeya* in cultivation and Mauritius has several native species, so I suspected the ones in the nursery were hybrids. I asked if I could take a couple of seedling plants back to Kew and get them to flower to see if they came from two different species. If we could get someone from the Jodrell Laboratory at Kew to run a DNA test, that would dispel doubts even more quickly. The purity of the plant is important because it very rarely sets seed. If these seedlings did belong to the true species, we could cross-pollinate them to produce more, and that would be important for the future of the plant. If they turned out to be hybrids, we would have to remove them from the breeding line.

The second time I visited Mauritius they said, "Guess what, Carlos? Vincent Florens and Claudia Baider from the herbarium have found another *Dombeya mauritiana*, but it's in a totally different place from the original plant." It was inside a massive thicket of strawberry guava (*Psidium cattleianum*), an introduced invasive species that creates havoc but has nice fruits—like a mix of strawberry and pear, with a zingy punch. (Incidentally, eating them causes problems if you want to visit Round Island. Because the island is a nature reserve, equipment has to be frozen to avoid introducing unwanted animals or other organisms. There are no toilets there, either, and before you visit you have to stop eating fruit and vegetables for a week. Don't forget, we are seed dispersers too. Everything has to be quarantined, even your gut.)

This *Dombeya mauritiana* was in the middle of nowhere, at one of the highest points of the island in the Black River Gorges National Park, around two hours' walk from the road. To reach it, we followed a path into the mountains. Up and down, following the contours, over rocks and small rivers, hacking and chopping at the strawberry guava thicket with our machetes as we went. The thicket was particularly dense—people cut stems from the plant to use as canes to support tomatoes and peas in the garden, but the condition is that when you cut them you have to apply herbicide to the stump. (When I was first in Mauritius I assumed that red hands were some kind of traditional custom;

up in the highlands I learned that the herbicide contains a red dye.) It was a long, hot, uncomfortable walk.

We had been trekking for more than an hour, the silence broken only by the swish of blades, when suddenly there was a shout from the front: "Ganja, ganja, ganja!" They had discovered a cannabis plantation hidden in the forest. Their reaction was over the top: they tried to block my view and held up their hands, shouting, "Don't look, don't look!" as if they were shielding my eyes from a naked lady in a bath. "This is very serious," they said. "We will have to call the police and possibly cancel the trip." One of them reached into his backpack and brought out a satellite phone; out poured a torrent of French. "Ganja, ganja, ganja" was all I could understand.

Despite their fears, the trek continued: hack, chop, hack, chop, through a seemingly endless thicket of strawberry guava until finally, two and a half hours later, we scrambled up a steep hill toward the top. There, in the middle of nowhere, was our spindly stemmed specimen, about fifteen feet tall, topped with a tuft of branches and with leaves of about six inches, much larger than those I had seen previously. It was *Dombeya mauritiana* in all its glory.

I was under the impression it was going to be like the specimens at Kew—a low shrub with multiple branches—but the branches were out of reach, high above the guava thicket.

"We need a ladder," I grumbled, as we looked up into the tree.

"Oh yeah, probably," someone said, unhelpfully.

We wanted to get stems for cuttings, and blooms and leaves for herbarium specimens, and we also wanted to see if the tree had any fruit, all of which was impossible to do from the ground. Going back to get a ladder was not an option—I was flying home the following day—so how were we going to get our material?

The answer was a human ladder, with the stockiest guy at the bottom, two others standing on top of him, and me at the top. The problem was that below us there was a 300-foot drop into the valley. If anyone had slipped, that might have been the end of us all.

The two men stood on the shoulders of the base man, who strained and straightened with a giant roar like a weightlifter. Then, treading on shoulders, hair, and ears and apologizing profusely as I went, I scrambled and wriggled my way to the top of the pile, eased myself upright, and reached out as far as I could with outstretched fingers to try and grab the lower branches.

I was still a foot short.

A shout came from below.

There, bobbing about on the branches of a nearby tree, was a Mauritius bulbul (*Hypsipetes olivaceus*), a rare endemic songbird with a population of only about 290 pairs. As I reached for my camera, the human ladder wobbled alarmingly for a split second.

We decided to have another go at reaching for the lower branches. This time one of the rungs of our human ladder grasped a forked stick nearby, broke it off, and passed it up to the top. I hooked it over the branch that I wanted and snapped it off. The flowers were all male, exact replicas of the one in cultivation but double the size. And there was a fruit. It was so exciting to be able to preserve this particular clone as well as the fruit. But it was also confusing. The general consensus was that some trees of this species had male flowers and some had female. The flowers on this tree were all male. Yet a fruit, and seed, proved that this tree was . . . female?

We had been growing this species at Kew in the tropical zone. But having found this new clone in a temperate habitat, when I returned I decided to put some of the plants from both the lowlands and highlands in the temperate and tropical zones. I was convinced something temperature-related was influencing the flowers, just like the café marron. Perhaps temperature would do the trick?

Then something amazing happened.

In cool winter temperatures, the flowers from both the clones were female. In higher temperatures, the flowers from the clones were male. So, in the tropics, where it is hot, both are always male; and in temperate zones, the flowers change from being female in the spring to all male as temperatures rise and summer nears, sometimes with an intermediate stage.

If you self-pollinate the flowers by taking pollen from a male flower in the tropical zone and placing it into a female in the cooler temperate zone, however, no seed sets. They probably need pollen from another clone for pollination to take place. What I have yet to do is pollinate a male flower from the clone in the highlands with a female from the lowlands, or vice versa, to see if this works. But both flower at different times, so we would need to collect the pollen and store it until the plant was ready. It is an adventure waiting to happen.

Mauritius has a cool, drier season from September to November, at the start of the summer, and at altitude it gets even colder; if the plant were more widespread, we might have an interesting scenario. Flies would be able to pollinate plants that are a long distance apart—so far apart, in fact, they might experience different climates on any given day. What an ecological quirk: different climates on different sides of mountains at different altitudes might just lead to different clones.

The only thing I can confirm from personal observation is that in *Dombeya*, low temperatures seem to trigger female flowers. There is also a tendency to have clones that seem male most of the time, but every now and then they surprise us with fruit.

I have never known another plant in which the temperature affects the gender of the flowers in this way. We all thought that it was a lowland plant, but now we have a highland plant with a somewhat varied form, suggesting that at some point they probably covered the whole island and that there were once lots of different forms.

The downside of having just two or three survivors is that we have lost a huge portion of the big picture. Conserving plants in Mauritius is like archaeology, and trying to fit all the pieces together is complicated, to say the least.

I badly needed to discover if the plants in the nursery were hybrids. So we studied the DNA of those with different-colored hairs, and guess what? They are all the same species.

With the plant of the lowland setting seed, and the one in the highlands at Black River Gorges with fruits and male flowers, perhaps there is a true *Dombeya mauritiana* somewhere else that is pollinating these plants, one that we have yet to find. We know the location of most endangered plants on Mauritius; if there is another, we desperately need to find it, or the plant might be chopped down or die.

By the sounds of it, we need a plant sexual health clinic in Mauritius. It feels like a new job description should be created: plant sexologist. As they say, "We are here to help, not to judge."

❊❊❊

Elaeocarpus bojeri, the lace tree, is one of the most beautiful trees in the world, producing masses of small, white, fringed, bell-like flowers. In 2010, the last two surviving *Elaeocarpus bojeri* were found near the most famous Hindu temple on the island, a kind of Mauritian Taj Mahal, next to a crater lake called Grand Bassin, almost 2,000 feet above sea level. At one time there were three trees left in the wild, two growing close together and one a bit farther away, but it was feared the single tree might succumb to ongoing construction and development in the area, so it was lifted and transplanted into the Forestry Service's nursery, where it was flowering when I visited the island.

The people from the National Parks and Conservation Service kept telling me that cuttings from this plant didn't work and the seeds didn't germinate, but the Service's native plant nursery had four small plants, so it must have been propagated somehow.

"Hang on, you are telling me you have four, but in the next breath you're saying that you can't propagate it. What's going on?" I asked.

I wanted to take a cutting, but they kept on stalling. First I was told the plant in the nursery was too small, then that those in the wild were too far away and it would waste time because it would take me away from more important things and success was unlikely anyway. I had to pressure them but could not understand why. It seemed like *Hyophorbe amaricaulis* all over again.

They changed their minds after I spoke to Kevin Ruhomaun, who worked for the National Parks and Conservation Service. He presented me with a challenge. "It would be great if you could find out how to germinate the seeds—we have lots of them but just don't know how to germinate them."

This tree was so beautiful, I just had to help it survive. When you fall in love, there is nothing you can do about it.

I managed to get someone to take me to the Forestry Service's nursery, so I could see the plant for myself. When I saw the "bois dentelle" (the local name for *Eleaocarpus bojeri*), it was flowering, all white and frilly, and the ground around it was covered with rotting fruit. I looked at the seed, and it was obvious that it was going to need "nicking," since it had a thick, woody outer cover. This meant I would have to chip the seed coat with a sharp knife, or rub it with sandpaper, to allow water to reach the softer tissue inside the seed and trigger germination. I wondered how the seed coats were broken in the wild. Would tortoises have eaten the long green fruits (which looked like sweet peppers) and the seed have softened when it passed through their gut?

I went to see the head of forestry and suggested that they try nicking the seed.

"Oh yes," he said, "We tried it once and put them in an acidic compost, kept them moist, then sowed two or three seeds and one germinated. I suggest you try that."

I thought to myself, "If you know this, why is nobody doing it, then?"

Later during my visit, just before I left for home, I went to the nursery at Curepipe to have a look at another specimen of the plant that I discovered they had there. The grass had been allowed to grow right up to the base, rather than leaving a circle of weed-free soil to protect the plant from mowers and strimmers, and the grass had been freshly mown right up to the trunk. On seeing it, I panicked and ran up to the base of the tree, pulled away the clippings, and found that the strimmer cable had girdled the bark. I let off a salvo of expletives. Take away the bark and the layers below and you cut off the food supply between the roots and the rest of the tree. Water may still go up, making the tree look lush for a while, but chances are it will die.

I approached the head of the nursery.

"You see this tree? Bad news—it may look happy, but this tree is technically dead. What has happened?"

"Oh, you want to take cuttings?" he asked.

"No, I want to tell you that this tree is technically dead. It may reshoot from the roots, but the top will soon be dead."

I showed him the damage and he was equally frustrated. "Death by strimmer" is a common demise for young trees almost anywhere in the world, even in the UK, as any park manager will tell you. It's always a good idea to keep a small circle of clear soil around a tree, and perhaps even put a small cage around it if it's planted in grass. If not, edging "just a little closer" to make things tidier often proves fatal. Trees, especially young ones, dislike grasses and other herbaceous plants growing close to their stems and trunks in any case, as they have to compete for water, so the exclusion area is a win-win.

In the end, I traveled back with the seeds and cuttings I harvested from the tree in the Forestry Service nursery, rooted the cuttings, and germinated the seeds. It was as simple as that. The tree was growing well at Kew, it was exciting, and there were different clones, so I could cross-pollinate them and get plenty of variation; now there was nothing to be done other than learn how best to cultivate the tree and make it and my knowledge available to others. It is beautiful, it's critically endangered, and I can propagate it. A win-win-*win* situation. It even starred on the cover of the International Union for the Conservation of Nature's red list of endangered species, and became something of an icon of conservation.

Three or four years later, I got an email from Kevin Ruhomaun, saying, "Are you still growing this plant? If so, how many do you have and can you send us some?" Then I received a similar email from the nursery at Curepipe.

"What the heck is going on?" I thought.

I replied: "Yes, I have five seedlings and three or four plants grown from the cuttings taken at the Forestry Service's nursery. I can send you some cuttings. How many do you want?" And, more important, "Why do you want them?"

Someone who was more concerned about general landscape

than conservation had cleared the vegetation to improve the view of the lake. In the process they cut down the two Elaeocarpus trees next to the Hindu temple without even realizing it. When the news broke, the international plant conservation community was shocked because they knew how important these trees were.

I sent the nurseries instructions telling them how to germinate the seeds and take cuttings—now they could do it themselves. After a while, the stumps in the wild started re-sprouting from the base. They are likely to have multistemmed trunks and might lose their elegance, but at least they have survived and, in time, may recover their full splendor.

The battle between those who use propagating knives to take cuttings and those who use axes to fell trees is like the one between David and Goliath. But we know who won. Indeed, we have more lace trees at Kew now than could be felled on the whole island. The cutting edge of conservation is truly a knife blade.

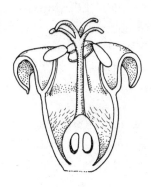

7

The Recycling Plant

Conservation starts in the most unusual places. Some projects end up in the trash; others, like this one, *begin* in the trash.

One day in 2007, while I was working in the nursery at Kew, I went to put some compost into a recycling skip and there, among the old leaves and vegetation, were some copies of *Curtis's Botanical Magazine,* dumped there while some vacant offices were being cleared out. The person who dumped them obviously didn't know what they were. The magazine is all about plants and is renowned for its botanical illustrations. It has been published by Kew for centuries and is highly respected.

I couldn't believe they were being composted. "I'm having those," I thought.

They were a bit mucky but undamaged, so I picked them out one by one and carefully brushed off the compost, revealing the artwork on the covers underneath. I'd cleaned up several when I recognized the unmistakable blooms of *Hibiscus fragilis* gracing the cover of volume 13, from November 1996. Flicking through the pages, I found the whole volume was devoted to the flora of the Mascarene Islands. It covered many different aspects, from the history of botanical exploration to vegetation and conservation, and there were studies of particular plants—among them *Ramosmania rodriguesi* and *Hyophorbe lagenicaulis*. I realized I had read some of the articles individually online but had not seen the whole issue before. On page 200 was a plant called *Lobelia vagans,* which was a new species to me. It looked for all the world like the blue lobelia you could buy from the garden center, its soft tumbling foliage covered with masses of delicate, finely cut flowers, except that in this case the flower was white.

On reading more, I discovered that *Lobelia vagans* was an endangered local endemic found only on Rodrigues and nowhere else in the world. It had been collected by Isaac Bayley Balfour in 1874 and later by Jean-Yves Lesouëf. Lesouëf was so passionate about saving endangered plants that he took his family to the island on holiday, where he stumbled upon this particular species. He collected seeds and brought them back to Brest Botanic Gardens, where he successfully raised some plants, then donated some of them to Kew. The article went on to explain the conditions in which Martin Staniforth, who was in charge of the Temperate Nursery at the time, had displayed them in hanging baskets in the Waterlily House.

I had never seen this plant in Kew's living collections, and I had the sinking feeling that we had lost it.

That evening at home I read the article more carefully, absorbing every detail. I needed to know more about it and what had happened to that plant. My next stop would be the database of plant records at Kew, where every plant that has ever been grown in the gardens is recorded. Each plant is identified by a number. Even if the plant name changes (as it often does), the number always remains the same. We even record deaths in the searches.

I arrived at work early the following morning and headed straight to the database to find out what had happened to *Lobelia vagans.* After all, it may still have been growing in the Palm

House and I just hadn't seen it. Kew is such an endless resource of hidden gems that this occasionally happens.

When I searched the database, it came up with the message "seed store," and I was relieved—even though we were not growing the plant, we still had some seeds. Lobelia seeds could last for ages, so these should still be all right, I thought. I could understand why we were not growing it: it's a high-maintenance plant and needs propagating regularly—it is not like a tree or shrub that you put into the ground and it will last for decades. If you grow this lobelia for several generations, you eventually have problems with inbreeding too.

Discovering this plant in the magazine was like a message from the past. I knew that nothing had been done with it for more than a decade and that it was time to grow it again, create more generations, and produce a batch of fresh seeds to ensure its survival.

I went down to the nursery's seed bank, where millions of seeds are kept in alphabetically arranged bags inside sealed boxes, each bag with a packet of silica gel to keep the seeds dry. A, B, C . . . I ran my fingertips over the bags until, toward the back of the cabinet, I came to *L* for *lobelia* and, finally, *V* for *vagans*, written on a scrap of fading paper. I opened the bag and inside was a brown paper envelope. Excitedly anticipating a small pile of genetic gold dust, I peeled back the flap, upended the envelope to shake out the seed, and . . . nothing happened.

It was empty.

I could not believe it. My heart started to race—there must be some seed in there somewhere. I needed only a few. I pulled out my knife and carefully sliced open the envelope to see if there were any trapped in the corners.

There was absolutely nothing. Not one single seed.

A horrifying thought washed over me: "What if this plant

became extinct in the wild because it had not been collected for many years, and we didn't have any seed?" It would be too much to bear.

I was just about to throw the envelope away, when yet another thought crossed my panicked mind.

The envelope was one of those old-fashioned "lick and stick" types, with a line of glue along the inside edge of the flap. What if, by some outlandish chance, some of the tiny seed had stuck to the glue while the envelope was being sealed?

I carefully peeled back the strip and, to my joy and relief, my hunch was right. Some of the tiny seeds were indeed stuck to the glue.

But what should I do next?

There was no way they could be separated from the glue, so I cut off the flap of the envelope that held them—a strip about two inches long and one half inch wide—and laid it on some moist compost so the brown paper could absorb the moisture, then put them in a humid propagator to germinate.

I ended up with about fifteen new plants along that little strip of glue.

It just shows you how incredible plants can be, and the fine (and sometimes weird) margins between preservation and extinction; in this case, the difference was a magazine dumped in the trash and a single lick of an envelope.

I grew the plant, and it flowered and set thousands of seeds. Some were sent to the seed bank to be stored for the future, and I kept propagating the plant from cuttings, to avoid further inbreeding. We now have some of these treasures in cultivation.

The plant needed to go back home, back to where it belonged. On my second trip to Rodrigues, I asked the local people if they had seen it. They said they hadn't, though they told me that they knew it was once found on the island, in Cascade Mourouk, Cascade Victorie, and Cascade St. Louis, and nowhere else in the world. Apparently they had looked several times and couldn't find it.

The law on Rodrigues means it is impossible to import potting composts to the island, so the nursery staff dig soil from the surrounding habitats and use that instead. The problem is that it is full of seeds from invasive weeds, which germinate easily and swamp the seedlings of the native plants.

To fix the problem, I went to a local shop, bought a microwave oven, and spent an afternoon sterilizing trays of soil. It killed every single seed apart from one, *Leucaena leucocephala,* which was quite a feat of survival because the compost was piping hot. If it could survive fifteen minutes in heavy clay at boiling point, it would probably survive a nuclear explosion.

This technique is used only when sowing and germinating rare seeds, not for large quantities, because the microwave is so small. Once the seedlings are large enough to look after themselves, they are transplanted into normal compost.

One day, when I was back at Kew, I sent some pictures to the staff from the Mauritian Wildlife Foundation nursery in Solitude, Rodrigues, just in case they stumbled upon it and needed something to help with identification. As far as I knew, it hadn't been collected in the wild for a long time, so there was a good chance it was extinct in the wild. Later, we managed to get the nursery staff from Solitude over to Kew so that they could be trained, and they sowed some seeds while in London.

In October 2016, I received an email from Alfred Bégué of the Mauritian Wildlife Foundation on Rodrigues, who said: "I think we have found something interesting here. I wonder if you think it is the plant I am thinking of? We have found it growing in soil collected in Cascade Mourouk, from a river deposit, and the plant seems to have come as a weed in the compost. Can you please confirm its identity?"

The second I read that, I thought it must be *Lobelia vagans*, given the location, the type of habitat, and the description of the compost, but quickly changed my mind. Surely it was too much of a coincidence . . .

I clicked open the picture. There, on my screen, was a small flowering *Lobelia vagans*, growing out of a nursery bag filled with soil, together with *Aloe lomatophylloides*, the plant they were trying to cultivate.

What were the chances? Just imagine, a seed floated in the breeze down the hill, in Cascade Mourouk, landed in the river, and settled into the sediment, until one day, just by chance, the nursery people dug into the soil. Then, with a glimpse of light and moist conditions, the seed germinated, grew, and flowered, revealing a species that had not been seen in the wild for decades. Perhaps there are plants up the hill and it floated down on the breeze quite recently; perhaps it had been lying there for years.

In conservation, sometimes all you need is a little bit of luck.

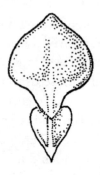

8

Water Babies

It is no surprise that after a few years at Kew my attention began drifting back toward my first love—waterlilies.

I was very young when I saw my first waterlilies. Perhaps it was the flowers growing out of the water, as if something magical was appearing out of nowhere. Perhaps it was their beauty, their scent, the mystery of those flowering at night, or maybe, if the Buddhists are right, I was a frog in another life. Whichever, I am a self-confessed waterlily addict.

My mother collected many plants: orchids, bromeliads, fruit trees, vegetables, flowers, and more. But no waterlilies. Then, while my father was developing our *finca*, he found a shallow underground spring. He hired a JCB to excavate a swimming pool–sized pit, where he could collect water for an irrigation reservoir. The soil, a mixture of conglomerate stone and heavy clay, was like natural concrete and almost waterproof; as if by magic, the pit filled up completely a few hours after he had dug it. My father also created a stream that fed into the reservoir and drained the excess water down the hill. All in one afternoon. The speed and scale of the operation blew my mind. Out of nothing a reservoir emerged, but it was just water—empty, sterile, and with no signs of life.

My imagination sparked to life. In my mind my eight-gallon

aquarium at home had suddenly become a thousand-gallon outdoor fish bowl. The discovery that other people's unwanted goldfish were hardy became my miracle of the loaves and fishes: after two years I could throw a piece of bread into the water and thousands of goldfish of every shape, size, and color would scoot up to the surface to eat it.

Another miracle was that so many species appeared naturally; insects, tadpoles, and aquatic plants all made it their home. But no waterlilies. They remained elusive. I just looked longingly at pictures in books and those that could break through the snowlike static on my black-and-white TV.

As far as I knew, there were no native or cultivated waterlily species in Asturias (though I later discovered that there is one very rare species in a well-hidden spot). Despite my parents' many contacts in the horticultural trade and among enthusiasts, no one seemed to have any for sale or exchange either. They asked everyone, without a sniff of success. I suppose Asturias has such steep terrain and fast-flowing rivers, there is no place for waterlilies to thrive, or much space to build ponds. So my first plant came from a very odd source.

There is an exhibition center in Gijón. It is regularly used for conferences and shows, but for two weeks a year there is a trade exhibition where you can buy anything, from smoked hams and sports cars to live chickens and leather sofas. One year, when I was about ten, a company used a garden with a pond containing waterlilies in their display. They instantly caught my eye. In the rush to leave, the exhibitors left behind an empty plastic pond and a pot with a waterlily. So I claimed the plant for myself.

I can't remember if I thought of this as recycling, rescuing, or stealing, but I had to explain its appearance to my parents. After justifying the theft by explaining its potential fate, I finally had a waterlily. Even though all the pads had dried, my new water baby had growth points that were alive, so we put it into a large water tank in the *finca* to begin with. My mother was worried that it would not survive in the pond.

I checked out my lily week after week. After the boring winter days, around late April and early May the pads grew and the fun began. Although there wasn't a label in the pot, I believe it was common *Nymphaea alba,* and it became one of my favorite pets. A couple of seasons later, I split the plant, threw it into the pond, and dozens of white blooms opened in midsummer—it covered the whole surface in no time.

Voila! Waterlily nirvana had been created in a few simple steps. It was like making an omelette, just in a thousand-gallon bowl. It was clear that deep down my mum loved them too.

I also learned something important: nature can be destroyed but it can also be created and transformed. Months later, when my brother Miguel was reciting Einstein's Law on the Conservation of Energy ("Energy can't be created or destroyed; it can only be changed from one form to another"), I said to myself, "Just like nature."

I didn't manage to get hold of another waterlily for years. But the group of plants that had caught my attention turned out to be complex and endlessly fascinating. I started gathering information from books and TV documentaries. I discovered that there were tropical species and that some had blue flowers (quite startling, when you have seen them in only white and pink), and I came across the eye-popping facts and figures about *Victoria amazonica*—the Queen of Amazonian nights, the giant waterlily.

※※※

How old are waterlilies? Well, their ancestors were already filling ponds more than 65 million years ago; half a million years *later*, the dinosaurs were toast. As some of the oldest flowering plants in existence, they still demonstrate some of the properties, pollination techniques, and form of early flowering plants. Sometimes we think of them as "primitive" because they don't have the complexity of an orchid, but that is unfair. Waterlilies still dominate the ecological niche they live in, and are still diversifying and adapting.

The color of their flowers can be dramatic too—they can be pink, blue, white, purple, or buttercup yellow, and some turn pink on the second day after opening. Some have a scent that smells of strong acetone, capable of transforming an Amazonian pond into something that smells like a nail bar; some are sweetly fragrant; and others have no smell at all. Although waterlilies are one of the earliest surviving forms of flowering plant, their flowers are nonetheless complex, with well-developed mechanisms for survival. They attract pollinators with their color and scent, they have accessible landing platforms, and they have sophisticated methods of seed dispersal just like highly evolved plants.

There are about seventy known species in the waterlily family (which is known as Nymphaeaceae). One of the genera in this family, *Nymphaea*, contains approximately sixty species, about fifty of which are in the tropics, mostly south of the equator. That said, they are cosmopolitan and can be found almost anywhere in the world except Antarctica and vast expanses of desert, so keep a lookout wherever you are. Some, like *Nymphaea alba*, are European, with related species in North America. Others, like *Nymphaea tetragona*, are found only in the freezing tundra of Canada and Russia. You'll come across waterlilies in the arid zones of northern Australia, which suffer flooding in the monsoon season, as well as coastal Peru and other unlikely arid and semi-arid areas where temporary and permanent watercourses are found. Eastern North America and Central and South America are the two areas where species diversity is at its greatest.

There are about twenty species of *Nymphaea* in South America. If you can't recall seeing pictures of South American waterlilies en masse in a "Monet meets Henri Rousseau with a salsa soundtrack" combo, it is because nearly all the South American species bloom at night. There are so few pictures of them on Google Images, you have to fire your imagination with botanical descriptions or a few herbarium specimens that show the corpse but not the spirit.

This is how the night flowerers work. On the first night the flowers open after sunset for a couple of hours and are female. The second night they turn male (or are both male and female), and some species will pollinate themselves, if the job has not been done already.

Their nocturnal lifestyle is one of the main reasons why they are not widely cultivated. Most gardeners give up and go to bed while the bud is halfway to glory. They look absolutely stunning and are well worth the wait, if you can stay up long enough, though they are better suited for an Ibiza nightclub or Nosferatu's second home in the Bahamas. Cultivating them is a handy way to understand the intricacies of their sexual habits. A few botanists I know brought their work home and studied them overnight while in their pajamas.

Tropical South America has compensated us with some waterlilies that provide daytime entertainment, including the incredible enormous pads of the giant waterlily, from the Amazon, truly a botanical wonder. There are also several species in Africa, some extinct in the wild, others known only in cultivation. All but one are day-flowering. However, the day-blooming claim to fame of the African species is outshone by the great diversity and beauty of tropical Australian species.

Australian waterlilies are quite simply extraordinary, the

kings of the *Nymphaea*s. There are twenty known species, most unique to Australia. They share the day-blooming habit with most of their African counterparts and, like them, they share their habitats with man-eating crocodiles, which makes collecting them a risky business. Add the many waterborne illnesses that you can get in large areas of the tropics and you can be sure that collecting from the wild will require not only permits but nerve and a strong heart.

✦

Mention Australia and most people think of kangaroos, Uluru (Ayers Rock), or a koala bear. I visualize monsoon storms flooding the plains, transforming the desert into a sea of waterlilies in white, pink, and blue, with charcoal-black swans paddling serenely past and flocks of snow-white cockatoos making their last passes overhead before settling in their roosts at sunset.

Some tropical waterlilies survive the dry season in a state of dormancy, then, as soon as the rains hit the watercourses, they start growing again. In northern Australia, billabongs, water holes, and creeks may be temporary, filled for just a few weeks a year. The soil can be brackish (slightly salty) and alkaline, or acidic and nutrient-starved. Despite this, plenty of plants survive in these sorts of places—the waterlilies that grow in these conditions are a subgroup of *Nymphaea* called *Anecphya*, and are found in the north of the continent: Kimberley, Northern Territory, and Queensland. Like all waterlily species, their feeding roots go deep into the mud below the water holes, lakes, and creeks. Once they have matured and stopped putting on new leaves, all their energy goes into fattening up the tubers, as they get ready to survive the coming drought. They also develop contractile roots: as the water evaporates and disappears, the large roots slowly retract, pulling the growth tip of the waterlily deep into the soft mud below, protecting it both from desiccation and from grazers and munchers on the lookout for an easy meal. They bury themselves deeply—I have seen them sink fifteen to twenty inches to the bottom of a large pot, and I often wonder how deep they would go in the wild. Seven to ten feet?

When I went plant hunting in Queensland, I visited a huge area that two months earlier had been a lake, ten feet deep, covered in fragrant blue waterlilies. But as we approached I saw it had evaporated and been replaced by what looked like acacia, eucalyptus, and long grass. There was no sign of the lake or waterlilies at all.

It was surreal to think that hidden under the grassland and trees were thousands of waterlilies, waiting for the rains that come almost every year. I say "almost" because although they have an annual wet season, the level of precipitation fluctuates wildly. Sometimes it can be a deluge of biblical proportions that floods thousands of acres of land; at other times it is dramatically less, or doesn't come at all. The dry season has unpredictable rainfall too, in patterns that could be termed "mosaic storms," falling randomly on small areas as they pass through the continent and leaving other areas untouched. These episodes have forced waterlilies to adapt or die. They have clearly done the former. If all else fails they will have dispersed hundreds if not thousands of seeds. Upon being released from the fruit, they have a device called an "aril," which is coated with air pockets that will float them away for a day or two, and then they will sink to the bottom of the pond. If the seeds are dry, they will also float for a while, making it likely that they will be dispersed once the rains finally come, only to sink again after a few hours of being in the water, germinate, and grow. The seeds are also sticky when wet, so if one adheres to a migrating duck or goose it may be transported to a new location far away and carry on surviving and colonizing. This is how waterlilies effectively colonize a whole lake or a new wetland area.

To cope with this uncertain existence, the seeds of some species germinate at any time, providing they are wet. Others produce batches of seeds that seem programed to wake up at various times, some the first time they are flooded, others the third, some up to nine seasons later or more. Some species produce tiny seeds that travel more easily and for a greater length of time and produce smaller seedlings; others have larger seeds that allow them to grow in deeper water. The lottery of geographical locations created by this odd dispersal and the daily, monthly, and yearly challenges they face have translated into a glorious diversity of species with the adaptations they need to survive.

All *Anecphya* species have one thing in common: they are pollinated by bees. What you see today if you visit these wild populations is a momentary snapshot of evolution, driven by weather fronts, wildfowl migratory patterns, and bees, morphing into different shapes and strategies, yet remaining 100 percent waterlily. Waterlilies have survived against the odds since before the dinosaurs became extinct, awaiting the next challenge that life throws at them.

Earlier in my career, there were only two species of Australian waterlily in cultivation—*Nymphaea gigantea* and *Nymphaea violacea*. Both were pretty rare. Some new species were discovered at about the time I started working at the aquatic collections at Kew, and a number that were described many years ago but had been lost were re-collected. This just added to my excitement, and seeds for most species that were once cultivated became available once more.

At that time most experts thought that *Nymphaea gigantea*

and *Nymphaea violacea* were difficult to grow to their full potential in cultivation. Most broke their dormancy as and when they felt inclined, rather than when they were needed for display. Often it would be too early or too late to benefit from the high light levels and temperatures of summer needed for flowering. Most of the time they became dormant again without displaying a single bloom.

People said, "The other species are just too difficult to grow, so forget them."

I said, "Give me the seeds so I can experience the frustration for myself."

I was madly, obsessively in love with them, so I decided to discover what was needed to make them, and me, happy. I also knew that I had an advantage. Most of the growers were in subtropical yet seasonal areas like Florida, but the Australian species prefer an equatorial and constantly hot environment. Growing them in the artificial environments at Kew gave me a leg up. Light and temperature could be provided constantly and evenly. Once I'd worked out their requirements, I could control their responses accordingly to produce displays for the public, or stimulate growth for DNA samples.

This was a story with a happy ending. If the right material is available at the right time, with the right paperwork and access to Kew's waterlily tanks (kept at 90°F, these are hidden behind the scenes in the nursery—we often joke about turning them into Jacuzzis when it is very cold outside), I have the facilities to grow every single Australian species that is available. I have not sourced all of them yet and am still missing three. But there is so much variation in color, size, and performance within individual species, plus the fact that a few types are still refusing to germinate reliably or produce more than half a dozen blooms before going dormant, that the fun is far from over.

So far, using techniques such as carefully drying out the dormant tubers to initiate growth and potting them at the right stage, I have put several elusive Australian natives on public display, and the feedback from amateurs, connoisseurs, and other waterlily "stalkers" has been stellar. I have also increased the number of species in cultivation and created a few new flamboyant hybrids, already coveted by tropical water gardeners around the world.

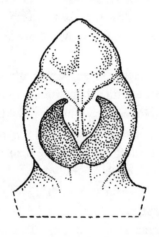

9

Victoria's Secrets

A tale is told of the beautiful young women of a tribe who in times past sat at night on the banks of the River Amazon, to sing together and dream of a better future, entranced by the pale light of the moon and the magic of the stars beyond. They believed that if they could touch the moon, or stars, they would become one with them. One night, as the rich organic fragrance of the tropical night intensified their desires and moonbeams shone softly through the trees, Naia, the youngest of the tribe and a dreamer, decided to climb a tree to try to touch the moon. But she didn't succeed. The next day, she and her friends climbed the distant hills. Perhaps she could reach it from these? But, alas, again she failed; the moon was just too high.

The following night, Naia left the longhouse for a third time, finally hoping to fulfill her dream. She followed the track down to the river, and there she saw the full moon, radiant and beautiful on the water. In her innocence, Naia thought the moon had come to bathe in the river, and, seizing her chance, dived into the deep waters to touch it, only to disappear forever. Feeling pity for the plight of the innocent woman, the moon transformed Naia into a giant waterlily with an intoxicating perfume—the *vitória-régia* (now known as *Victoria amazonica*)—and she became immortalized in its beauty.

This legend among the Tupi-Guarani–speaking people in South America encapsulates the allure of this beguiling plant. Its majesty, the enormous size of its flat, circular leaves, and the prestige of growing it have inspired gardeners for centuries, and it will remain an icon of the plant world. When the first seeds arrived in Britain, there was a race between Sir Joseph Paxton, head gardener at Chatsworth House in Derbyshire, and Sir Joseph Dalton Hooker at Kew to be the first to get the plant to flower. Paxton succeeded and the flower was cut and sent to Queen Victoria.

The giant waterlily flowers at night and is pollinated by beetles. A first-night flower is white and female; pollen-carrying beetles arrive, are trapped all night in the flower, and pollination takes place. On the second night, the pollinated flowers turn pink and become male; pink is a color that the beetles can't see particularly well in the dark, so they don't return to flowers that have already been pollinated. That night, the beetles escape and fly off, dusted with pollen, to a new white flower, where the process starts all over again.

Every part of the plant, other than the petals and the upper part of the leaves, is heavily armed with sharp spines, which makes it look reptilian. The leaves lie on the surface of the water and have been recorded as measuring more than nine feet in diameter a number of times, though La Rinconada, a leisure park near Santa Cruz in Bolivia, once grew a plant with a world-record leaf that was a whopping 10.5 feet wide. That's more than 26 square feet of leaf surface. When you think that a single plant can produce six to eight of these pads, you can see how it can cover a huge area. We have had plants covering the whole surface area of the pond in the Waterlily House at Kew, which is 26 feet in diameter.

Slice a cross section of the flower and the inside looks quite bizarre. There is something almost alien about it. You would never guess that you can make something as benign as popcorn from *Victoria amazonica* seeds. Inside the flower there's a large empty chamber; the female receptive parts are at the base, and at the top of the chamber are some hard, fleshy, infertile modified stamens—botanists call them staminodes. Above them are the fertile stamens, carrying pollen, and then there is another layer of staminodes. This "sandwich" creates a cage above and below the stamens, which opens or closes, trapping or releasing the beetles, depending on the flower's stage of development and maturity.

❖

One way to learn about a subject is to read everything that's been published. It's not as easy as it sounds. Not everything we know is written down, and scientific papers themselves are not always easily accessible (sometimes you don't even know what's been published). The best way to learn, then, is to see the plants for yourself—ideally in their native habitats, to put them in context—and draw your own conclusions.

Years ago I read a paper on *Victoria amazonica* pollination published in 1975 by a former director of Kew, Professor Sir Ghillean T. Prance.[5] It was fascinating. Prance not only grew the plants in Kew's display greenhouses for his own research, but also visited the Amazon to observe them in their native habitat for himself.

As part of his experimentation in Brazil and Kew, he pushed a thermometer down through the middle of a first-night flower, past the stamens and into the chamber below, and discovered that this part of the flower generates heat. He tried again, with several flowers at the same stage of development. The temperature inside the chamber was the same inside them all: 90°F.

He deduced that if you were in the middle of a large open area that is cooler than the flowers, and the fragrance was warmed, it would vaporize, rise up like an air balloon and drift farther away, increasing the chances of attracting pollinating beetles. He also specified that this happened only on day one.

I always felt that the vaporizing of the scent wasn't the whole story, though, because I had noticed a couple of things while tending the waterlilies at Kew. I knew that second-day flowers started to open earlier than first-day flowers but never opened completely before sunset, and I wondered why. To complicate matters, in the UK the sun sets at about four o'clock in November and after nine o'clock in June, and I wondered if this made a difference to the times the flowers opened. At Kew we grow just one or two plants a year, so we rarely have first-day and second-day flowers on the same plant, or flowers at the same stage at the same time (we would need many more plants for that to happen). From long-term observations of Kew's collection of day-blooming waterlilies, though, I'd also noticed that first-day flowers often open a little later and close a little earlier than they do on the second night and subsequent days.

I am rarely around Kew at night, not often enough to make sense of when things happen due to fluctuations in the daylight hours in the UK. We are also missing the beetles that are needed for pollination, so this is done by gardeners instead. We have the "fat lady" singing, but not the choir of tenors. I was fortunate, though, to continue my observations while out in the Amazon basin, when for a time I led groups up the Amazon on a cruise ship, sailing out of Iquitos in Peru, as the on-board naturalist and botanical expert. Unlike most of my rainforest experiences, the accommodation on this trip was luxurious, with soft carpets, air-conditioning, and massive breakfasts, but it doesn't matter how many luxuries there are on board the ship—life's still tough out in the rainforest.

Both river and rainforest were filled with wildlife, though the best time for viewings was between dusk and dawn. We saw rainforest curiosities like fish-eating bats, yacaré caiman (Amazonian crocodiles, which are quite small compared to the saltwater behemoths in Oz), and crazy, colorful frogs. But the most amazing sight of all was when, traveling along small tributaries by speedboat, we discovered vast areas of the rainforest illuminated solely by millions of fireflies, as if the tree branches had been densely covered in bright flashing lights. It was like Christmas in the Amazon.

During my trips up the river, we spotted more than 200 species of birds, as well as red howler monkeys, capybaras (the largest rodents in the world—they look like giant guinea pigs), and hoatzins (large birds like a pheasant with a spiky crest and a pale blue face). We fished for piranha using steak as bait, saw plants in abundance, and explored flooded and terra firma forests. Yet the most exciting experience of all was seeing *Victoria amazonica* in its native habitat.

To encounter *Victoria amazonica*, you have to leave the comfort of your boat, with that nice air-conditioning that blows the mosquitoes away, and trek through the hot, steaming jungle to find a lake. Most people think that the giant waterlily grows in the Amazon river, but it doesn't; it grows in the Amazon *basin*. The river meanders a lot; sometimes the curves are so pronounced that they meet at the top and join together, forming a straight line and leaving behind an oxbow lake. It is in these lakes that the giant waterlily is generally found.

We went out late one evening at the beginning of the dry season, just as the sun was setting, to see one for ourselves. After leaving the comfort of the boat, it was a long trek, through winding pathways with dense forest all around, but after an hour's walking we eventually arrived at the lake, in a clearing. A boardwalk led out into the lake, so we could get closer to the giant waterlilies, and there were about six or seven plants in flower toward the center of the lake, the nearest being about 15 feet away from the edge. Even from that distance, it was enough to help me understand what was happening. I could see a few plants with second-day (pink) flowers in full bloom but with their inner petals closed. The inner petals don't open until about sunset, or just after, when the buds on the first-day flowers are opening and the beetles have new flowers to fly to.

I told the group all about this glorious waterlily and how it was pollinated, and they admired its magnificence from afar. But what I was desperate to see was the only part of the pollination process we miss at Kew: the beetles. The local guides usually cut a flower at dusk to show everyone in the group how the waterlily beetles are trapped inside the chamber, before escaping and flying away. At last I was going to see this for myself. I was going to meet the stars of the show.

As the sun started to set and the sky became a blaze of orange, I noticed a species of caracara, a South American hawk a bit like a buzzard, standing on a giant waterlily pad, staring intently at the flower bud of a second-day bloom that was about to open. It was concentrating so hard, it didn't move. When I asked the local guide what it was doing, he told me that it was waiting for the beetles; their challenge was to come out of the flower and escape before being eaten by the bird. "Perhaps they all come out at once," I thought. "Then at least some will survive through sheer numbers."

The guests were thrilled to have seen the legendary *Victoria amazonica*, one of the great sights of the Amazon and the botanical world. But as the sun was about to dip beyond the horizon, suddenly a huge swarm of mosquitoes emerged, and, unwilling to wait for the guide to find a bloom close enough to the shore to cut, the guests decided to go back to the ship before darkness descended. As they hurried back to the main path and the safety of the boat, they had a parting request: "If you are lucky enough to find a flower that is close enough to cut, please can you bring it back to the ship?"

To me, this was the chance of a lifetime. I was desperate for the opportunity to see the beetles flying from flower to flower so that I could better recount the pollination story. But I knew we couldn't stay long; the ship wouldn't stay moored forever, so we were under pressure.

With almost everyone else on the way back already, the guide made one last effort to find a bloom within easy reach. He eventually gave up. "I think we are going to miss out this time," he said. The frustration and disappointment on my face must have been obvious.

There I was, in the Amazon rainforest, with a perfect specimen of *Victoria amazonica* right in front of me, barely 15 feet away from the edge of the boardwalk. This was my one and only chance, and I was unable to take it.

The water level in the lake was low. Between me and the plant was gooey, liquefied mud, the kind that looked as though it would suck you right under. Even if I survived that, who knew what was lurking out in the dark water beyond: piranha, pacu, electric eels?

Suddenly I heard another voice.

"What's the problem, mate?"

It was one of the guests, an upbeat guy who thrived on being resourceful and whose girlfriend had become fascinated by the waterlily. He'd heard me and the guide discussing the situation in Spanish, could see my frustration, and had worked out what was being said. He liked a challenge and decided to do something—what better way to impress his lady friend back at the boat?

"Do you have a knife?" he asked.

I did, together with the ever-present pruning saw in my rucksack, but before I could unfold the knife, the local guide handed him his machete. Cutting his way back into the forest, our new friend found two strong, straight stems lying on the ground and, after trimming them down to a similar length with the saw, he tied them to the saw and machete with a camera lead to make what was effectively a pair of giant, unwieldly loppers. With difficulty, we carefully inched the stems through our hands toward a second-day flower. After what seemed like an age we reached our prize and, with a final forceful jerk, pushed the blades on the stems together and, as steadily as we could, cut the flower. We watched nervously as it dropped perfectly into a loop our friend had tied onto the end of one of the stems, and eased the flower back to shore.

Its safe arrival was greeted with exultant shouts, applause, and whoops of relief. We gently placed the oversized flower in a sealed bag and put it carefully in my backpack, hoping there would be a few beetles inside. Just as we were about to leave, three macaws flew low over the water and headed off into the rainforest.

We ran as fast as we could back down the narrow path, now wrapped in the darkness of the jungle. We could only pray that the boat hadn't left without us. When we finally emerged from the rainforest, backpacks bouncing and dripping with perspiration, the boat was still there, its engine chugging impatiently. The ropes had been slipped from the moorings and it was on the verge of pulling away from the shore. We jumped, legs outstretched, and our feet hit the deck with a dull but comforting thud.

I headed back to my cabin to gather my thoughts. The minute the door was closed, the chugging of engines was replaced by a high-pitched buzz, like the sound of a nest of bees. I opened my backpack and there was a swarm of frantic beetles trying to escape and fly their way out of the plastic bag.

I realized that if I opened the bag, they would swarm out into my cabin before I could take a picture. There was only one option—put them in the fridge. The cool temperatures would slow down their metabolism to the point of inactivity. By the time they warmed up again, I would be able to show them to the group, take my photos, and put them out on deck so they could fly back to the rainforest and their waterlily habitat, their story properly told.

But there were two major problems. The bag was large and there was no fridge in my room. Someone had the bright idea that I should talk to the chef. There was nothing to lose.

"Chef," I said, "I have a weird request," and I explained what I wanted to do.

"No problem," he replied.

My immediate concern was that if one of the beetles escaped and ended up in a guest's salad, I would instantly lose my job—and the chef would too. So I put the sealed bag inside a sealed bag, then inside *another* sealed bag, then inside a Tupperware container for good measure, and popped it into the giant fridge before heading off to my cabin for a shower.

When I returned to collect my waterlily just before dinner, there, as I'd hoped, were the beetles, lethargic in the bottom of the bag.

After dinner came the opening ceremony. I took the flower from the bag, put it on the table, and carefully cut it open in front of the eager guests. Out of the flower crawled eight or nine beetles, just as they would in the wild. But what happened next was unexpected. I carefully sliced open the chamber and it was crammed with beetles; there wasn't space for even one more. If you took one or two out, you wouldn't have been able to see where they had been. I counted twenty-one in all, each the size of an acorn, in a 2.3-inch x 1.5-inch x 1.2-inch floral chamber. Everyone was transfixed.

In the ovary of the waterlily, where the seeds are formed, are the starchy segments. One theory is that the flower burns the starch to create heat; another is that the beetles eat them. Based on my observations, both theories appeared to be correct.

It was obvious from the damage that the segments had softened enough for the beetles to eat them, as if they were marshmallows. There were so many beetles stuffed in there, though, that they had also devoured something like a quarter of the ovary itself. At least one of the greedy insects was still stuck inside. Bear in mind these beetles are not the same kind of pollinators as bumblebees, which go from flower to flower and constantly move on; the beetles actually stay overnight and live part of their life cycle inside the flower, where they compete for mates and feed.

After about an hour and a half, the beetles started to move, slowly at first but gradually becoming more active. Soon enough, they were moving around quite happily, but they still weren't making any attempt to fly away. I knew they were sensitive to low temperatures, but this was puzzling.

Being pollinated by night-flying beetles that don't function in cool conditions didn't make much sense to me. Common sense says that the flower should release the beetles in the heat of the day, once the sun has warmed them up to the right temperature to fly. It is not that cold at night in the Amazon basin (usually 79–80°F, though temperatures can drop to 68°F), but it seemed to be enough to stop them from flying.

Professor Prance recorded a temperature of 90°F in the chamber, which became something of a magic temperature for me. It must have been at least this warm when they were flying around like crazy in my backpack before I stuffed them into the fridge.

Prance had also reported that only the first-day flowers warmed up. The beetles, however, flew from a second-day flower. What was triggering their release? I wondered whether it was because the beetles start to eat the ovary, just like the ones I had here. Perhaps the outer layers of second-day flowers open up earlier than first-day flowers to avoid the ovary being eaten completely.

This would go a long way to explain the protective "staminode sandwich." Besides trapping them inside the chamber, the barrier would also stop the beetles from reaching the stamens on the first night, so they didn't get eaten too. They have a job to do on the second night, when the beetles get covered in pollen on their way out to another flower. But then, why is there another layer of staminodes above them, and prickly bud cases too? Perhaps their job is to stop the caracara from damaging the flower and getting to the beetles inside? Or maybe the outer layers open up earlier to help warm the beetles, since they have their central heating cut after the first night? Maybe the eating of its "marshmallows" takes its toll on the plant's ability to generate heat?

So many questions. A piece of the story was missing.

Here we were, late at night. The beetles were moving, but not much. I was expecting them to fly; they didn't even try. I just kept them in the box outside my window, in a safe place where nothing could eat them. Soon after the sun rose the next day, there was a massed whirring sound as they began to move their wing cases at speed; one by one they flew from the open box on the balcony of my cabin, back toward the rainforest and their waterlily home.

When I returned to Lima a few days later, I searched the Internet to see if I could find anything about the beetle, or the flower temperatures—just something to explain the beetle's peculiar behavior.

I learned a crucial fact from a paper published in 2006: the beetles don't fly at temperatures below 90°F.[6] Prance had reported that the flowers warm up to 90°F on day one, but he had measured the temperatures inside the chamberlike structure, down by the ovary and stigma, where the beetles spend the day, night, and following day. New research using a thermo-camera had revealed that on the second day the flower warms up near the stamens.

Now I understood. The beetle's favorite temperature was 90°F. When the temperature in the chamber started to cool, they moved toward the stamens that were shedding pollen, where the 90°F heat was maintained, because the stamens were still covered by the central petals that were folded inward, trapping the heat within the flower (a bit like wearing a hat). And so the second-night action would begin. At about 6:30 p.m., the pink second-day blooms would be ready, their beetles hot to fly, primed with pollen and ready for action. Next would come the jailbreak, with the caracara picking off stragglers. The beetles would walk out of the bloom over the flattened "petals," like aircraft taking off from a carrier, and fly out into the night in search of a new, white-flowered female. Only visible to those that are close by, she vaporizes her perfume to attract distant beetles, to make doubly sure the job is done. When the beetles find a white flower, they are enticed down toward their favorite temperature inside the chamber and, covered with pollen from the previous night, they then pollinate the flower and the process starts all over again.

The fact that the ovary was eaten inspired another thought.

Prance had found that even partially eaten ovaries still set seed, just in smaller quantities. Once the flower is pollinated, it sinks below the surface of the water, evicting any beetles that have been left behind before they do anymore damage, and uses the warmth of the water to ripen the seeds.

I chuckle when I imagine these late-night beetles going from flower to flower, or club to club, living their life of Amazonian beetle hedonism. I can imagine them shouting to one another every evening as they fly: "Okay, guys, this party is over, where to next?" There are males and females trapped in there, eating their fill and mating.

Perhaps the giant Amazonian waterlily is not as "primitive" as it appears. It provides housing, protection, heating, food, aromatherapy, sex, and a social life—what more does a beetle need?

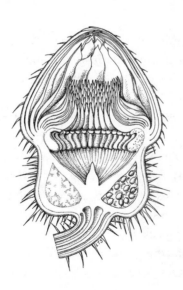

10

Warm Waterlilies

A passion for plant conservation. A long-term love affair with waterlilies. I began to wonder if there was a way of bringing them together. Was there a waterlily out there that was extinct—or critically endangered—that I could help to save? There certainly wasn't an iconic story like *Elaeocarpus bojeri* in Mauritius—but perhaps that was because no one had ever looked.

I started working my way through the catalog of waterlily species and discovered that *Nymphaea stuhlmannii* from Tanzania—one of only three yellow waterlily species in the world—was feared extinct. No one had collected herbarium specimens since the early twentieth century. Attempts had been made to find it many years ago, but with no luck.

Then there was *Nymphaea divaricata*, which is remarkable. It is one of only two waterlily species that don't grow floating pads—its leaves are underwater and shaped like a bow tie. It lives in rivers and dissipates the energy created by the current by swinging back and forth like the blades of a propeller. All you see from the shore are waterlily flowers popping out of nowhere. I have never seen a picture of it and don't know if there are many specimens about; I have seen them only in the herbarium at Kew (which demonstrates the importance of herbarium specimens). It was last seen about fifty years ago in Zambia and has not been seen since. As far as I'm aware, no one has specifically searched for *N. divaricata*, despite its history of being found in three countries—Angola, Zambia, and Zaire. Now we can use drones and satellites to help us, and surveillance photography is better than ever (Google Earth proves that the habitat where *N. divaricata* was last collected is still there), but information about them is still scarce. I would love to go and search for them one day. So far, collecting permits, bureaucracy, funding, and time have kept me from chasing this particular dream.

Then I read about another African waterlily, *Nymphaea thermarum,* a tiny species that had only recently been found, and in only one location. It is a waterlily that breaks the rules. It doesn't grow in a stream; it doesn't grow in a river; it doesn't even grow in a lake.

It grows in hot springs.

"I need to grow this," I thought.

Something was telling me that a single location for an aquatic plant was the perfect recipe for extinction. I asked colleagues at Kew and other experts around the world, and discovered that there were fifty plants left in the wild and two in cultivation, but no one knew how to propagate them. This was my chance. There would be no better waterlily with which to become obsessed.

In 1987, Professor Eberhard Fischer, then a twenty-five-year-old German undergraduate, was out in the Rwandan wild, researching the vegetation of the Albertine Rift. Bad luck dictated that his car broke down, but good luck meant he spent several days camping next to the hot springs called Mashyuza, in the Bugarama Plain, where he discovered a tiny waterlily. This was a waterlily that wanted to be found.

The hot springs were at the base of a limestone quarry, only a few miles from a cement factory, and the springs bubbled up into a large green pool. In the overflow, at the base of a small waterfall, where the water drains away, he found his tiny waterlily. The leaves were only about an inch in diameter, and the edges of the pad were smooth (in other tropical species, they are serrated). The whole plant was only between four and eight inches across. He immediately thought it could be a bit of a discovery. A year later the new species was named *Nymphaea thermarum*.

He took some plants back to the Botanic Garden of Johannes Gutenberg University Mainz and Bonn Botanic Gardens, where they were cultivated in greenhouses. Interestingly, the plant remained small even in cultivation, where growing conditions were better. Even more fascinating was the fact that it would grow happily in cooler water. It was originally found growing at the edge of the springs, where the water temperature was about 104°F. That was some feat of survival. After a painstaking search of more than fifty hot springs in the Albertine Rift, including those around Lake Albert, Lake Edward, and Lake Tanganyika, he did not find it again.

One small species, in one small location, is one vulnerable plant. I knew that there were plants at Mainz and Bonn, but it always helps to have duplicate collections. Even with expert cultivation, the more collections there are, the safer the plant is.

One day I received an email from the staff at Bonn asking if they could have some specimens of several endangered plants we have here at Kew, because they needed them for research. It was the perfect chance to ask for some seeds of *Nymphaeum thermarum* and other interesting and endangered plants.

By then, this waterlily had been growing for more than twenty years at Bonn and had produced plenty of seeds. "You can have as many *Nymphaea thermarum* seeds as you like," they said, "but be warned, the seeds germinate, produce the seed leaves, then die before they reach the surface." They had learned how to germinate the seeds, but no one had ever been able to work out how to grow the seedlings to maturity. I was immediately intrigued.

"It can't be impossible," I thought. "There has to be something that can be done."

The first batch of *Nymphaea thermarum* seeds arrived at Kew from Bonn in July 2009. I sowed them using standard methods and it was business as usual—they germinated and looked like blades of grass before producing the first seed leaves, a characteristic unique to waterlilies. What was the problem? Not long after, they stopped growing, began to look sickly, and, almost as quickly as they had appeared, they were gone.

Normally you propagate waterlilies by splitting the "roots" (they're actually rhizomes) into smaller sections to make new plants. Seeds have to be sown when they are dry, but when you do this, they float . . . so I fool them. I put a pot of compost in water, so the compost is at, or just below, the water level, then scatter the seeds on the top. After a while, usually overnight, they hydrate, then sink. In the morning I put a layer of sand over the top to hold them in place and leave them to germinate.

It is a bit like cooking. You have to have a recipe. It is not magic; it's logic. You can't just throw an egg into the frying pan as it is and expect delicious food to appear. What you have to do is crack the eggshell on the side of the pan, prise the shell apart, allow the yolk and albumen to slide gently down the side of the pan without breaking, and cook it for the correct length of time at the right heat to produce the perfect fried egg.

It's the same with waterlilies. You have to put the pot carefully into the water, lowering it slowly until it stands firmly on the bottom. If you drop the pot straight into the water, the compost will be washed over the rim of the pot and the seeds will be displaced. You also have to saturate the compost fully first, so that air bubbles don't run through the compost and move the seeds. I tried, failed, tried again, and kept modifying the technique until I was successful. When the pots are covered in four to six inches of water they produce submerged leaves a bit like a baby lettuce, until they are large enough to send out the first pad, which will grow up toward the surface. Most of the time it is quite easy.

I tried this with *Nymphaea thermarum,* but it just wouldn't take. I realized I needed to think laterally: to run through all the permutations, then experiment. I started to go through the factors that affect plant growth: temperature, the pH of the compost or water, the concentration of salts (which is usually but not always related to its acidity or alkalinity), and light (both intensity and length). If a different water temperature didn't work, perhaps we could try using tap water (it is quite alkaline at Kew), or the water filtered by reverse osmosis (which is almost like distilled water) that we use for the greenhouse plants. Then we could try a mix of peat and sand to pot it in, as a nice low-nutrient, acidic compost; or straight loam because it would be more alkaline and have more nutrients.

I experimented by sowing four or five seeds in different conditions (we had about 200, so there were plenty to work with). Most hardy waterlilies divide naturally, but some tropical waterlilies hardly ever do, and *Nymphaea thermarum* had hardly ever divided at Bonn in twenty years, so I knew that I couldn't rely on that. If we were to secure *Nymphaea thermarum* in cultivation, there was only one way: it had to be reproduced by seed.

Nothing had any effect. Every one of the seeds hung in there, looking sorry for itself for three or four weeks, then, as the seed ran out of food, gradually melted away into the water.

If some had died within twenty-four hours while others lasted a week, I could gradually work out what they didn't like, but I couldn't work anything out from these results. What on earth was going on?

For weeks I was preoccupied with the fate of *Nymphaea thermarum.* Night and day I couldn't get it out of my mind, racking my brain for a way to crack the code. I just couldn't believe that one day soon this plant would be extinct and there would be no more material to work with. I had to do something.

I turned to academic research and dug out some text about its history . . . which was in German. Luckily, Felix Merklinger, a student at Kew, translated the text, which read:

> A very rare and beautiful species. *Nymphaea thermarum* was only discovered in 1987 by Professor Dr. Eberhard Fischer and is so far only known from the hot springs (40°C) Mashyuza near Nyakabuye. In the overflow of the springs, *Nymphaea thermarum* grows at approximately 24–26°C water temperature.

So the plant was not growing in the superheated water as I'd thought, but in the run-off farther down the hill, away from the pool. There was plenty to ponder.

One evening I was at home cooking tortellini. As I stirred and the water bubbled, it came to me: the "Invisible Man." Yes, CO_2, which threatens to wipe out our civilization unless we change our ways, could be the magic ingredient that would save this species. CO_2 dissolves poorly in water, especially in a tank, where the reserves are usually depleted quickly. Some aquatic plants are challenging if not impossible to grow in cultivation unless you artificially increase the concentration of CO_2. That had never been the case with any other *Nymphaea* I had worked with—but this was no ordinary *Nymphaea.*

CO_2 is essential to any plant. It is a well-known fact from schooldays (unless you were daydreaming during biology) that light + water + CO_2 is the basic formula by which plants, using photosynthesis, produce sugars; and sugars, combined with the nitrogen, phosphorus, potassium, and other nutrients that are found in the soil and garden fertilizers, produce the complex substances they need to grow. We give them all of these but not carbon dioxide; we leave it to them to take it from the air, and simply assume they have enough.

Many aquatic plants are super-efficient at capturing CO_2 when there are low concentrations in the water. Many species of *Nymphaea* can cope with low levels while they are submerged. Perhaps *N. thermarum* was an exception. I remembered from the aquarium I had as a kid that CO_2 diffuses in water, but not in the same concentration, and it takes a long time to replenish. There are massive reserves of CO_2 in a large lake with a few plants, but not in a small tank, and demand increases as the plant grows, while the volume of CO_2 remains the same. At a critical point, when the concentration crashes below a certain threshold, the plant struggles and can no longer grow. Once the waterlily pads are at the surface they are fine, because their breathing pores are on the upper surface of the leaf and they have more CO_2 than they need, but when they are underwater . . .

Why hadn't I—or indeed anyone else—thought of this before? A lot of aquatic plants are known to need supplementary injections of CO_2, yet for *Nymphaea* growers this was off the radar.

If old habits die hard, rare species die soon.

I thought about my options. There is a system that you can buy to inject CO_2 into the water, but it would cost thousands of dollars, and complex health and safety regulations would have to be followed. It would be too expensive and the process would be cumbersome. The other problem was that I had no hard evidence to show Kew's management that this would be guaranteed to work.

There had to be another way, and this is when the old saying "If Mohammed will not go to the mountain, the mountain must come to Mohammed" came in handy. Instead of putting the carbon dioxide into the water, why not put the waterlily in the air? Surely I could get away with that. My thinking was that if I put the plant in a quarter-inch of water and the leaves were a half-inch tall, they would be above the surface from day one and be able to take in carbon dioxide.

I re-sowed the seeds—some in a pot in water, with the compost one millimeter below the surface of the water; others in damp but totally soaked, undrained compost, with the seeds just resting on the surface of the compost. I suspected that if those on damp loam dried out even for a moment they would turn crispy, so I placed them inside a mist unit, which maintained 100 percent humidity. It worked. The ones that were a millimeter below the surface of the water started to grow their typical narrow leaves (they don't produce pads until later); those growing in loam did the same, but as they were not submerged, the leaves were smaller and thicker. Amazingly for those in damp loam, the first leaf was upright—it looked like a blade of grass.

I decided to try just one more thing—I transplanted a few seedlings from both locations into damp loam in a pot and put it in a container filled with water to the very same level as the loam in the pot. I then put it on a heating mat at 75°F in a bright location, much brighter than my previous experiments, and waited to see what happened. Amazingly, the plants grew even better than they had before.

Everything changed. Within two weeks I could see a dramatic improvement; about a month later, a proper lily pad appeared. They were growing like every other waterlily.

"Problem solved," I thought. I tried it again and again, building up the depth of water as the plants grew stronger, until the leaves matured and floated on the surface of the water. Over time the plants flowered and I built up a stock of seed. It is very difficult to say how much time I put into this; perhaps in my dreams, in the REM phase, I was still thinking about this waterlily. Obsession was the key, and my curiosity made it grow.

I decided to write up what I had done in an article—quickly, in case I was knocked over by the 65 bus from Kew to Kingston. I wanted to avoid what had happened in Bonn—I had been told that a horticulturist there had retired without sharing the secrets of how to grow the plant. This was the kind of thing that everyone needed to know to ensure the survival of the species. I wrote it up in the *International Waterlily and Water Gardening Society* magazine, as if it were a cooking recipe. Respect to the tortellini.

CARLOS'S COOKBOOK

Nymphaea thermarum recipe

Find a small container that can hold water. Look for a pot that will fit into it, smaller in width and just a little bit lower than the container.

Fill the container with water. Fill the pot with fine soil to the very top. Place the pot inside the container. The water level of container and pot should be exactly the same. Once the soil is totally damp and settled, sow a few seeds by sprinkling them onto the surface.

Remember the water level is crucial: it can go down for 1–2mm or up for 0.5mm, but air exposure is a must. If you have a tiny watering can, you can check and refill the larger water tank every day, slowly. Keep at 22–26°C. You can have water heating in a larger container and then place it in the smaller one, or place the small container on a heating mat or bench at 24–26°C. I have done both and both work.

The first filiform leaf will emerge, get out of the water and it will be a happy guy. Then it will get CO_2 from the air while being totally hydrated—the base will still be submerged in water. It will grow second and third leaves which are round in shape, and they will come out of the water or have the underside touching the soggy loam with the upper exposed to the air. They will sort themselves out if you are accurate with the water level.

Put the plants in your sunniest location.

A few weeks or months later, they should start to bloom. Separate them into individual pots when they are large enough to handle (when they have five leaves that are 0.5cm wide).

By the time I sent the magazine my recipe, the plants were about two or three inches in diameter and nicely established. As a special species with a fascinating backstory, it was ideal for an article in the highly esteemed *Curtis's Botanical Magazine*. Because the magazine is small, the leaves of most waterlilies are normally too big to illustrate as a life-sized botanical painting, but *Nymphaea thermarum* would fit perfectly. To my delight, the article was accepted and Lucy T. Smith, a botanical artist, was asked to illustrate the plant. She sketched it in the nursery, then took the sketch to the herbarium to complete it in watercolor.

As she was working away, serendipity struck. Professor Fischer, from Bonn, happened to be in the herbarium, and walked past Lucy as she was painting. Like any curious person, he looked over her shoulder to see what she was doing and when he saw what she was painting, he was stunned.

"Excuse me," he said. "Where did you get that plant material from, of the plant that you are painting?"

She replied, "Oh, there is a Spanish guy in the Tropical Nursery who has a hundred of them growing in there."

"A hundred?! *Nymphaea thermarum* is extinct in the wild. It has finished, expired, gone. The local people dug a canal so they could use the free supply of hot water for washing. Because of that, the hot pool dried out and the waterlily faded away. I thought the only living plants were in Germany," he explained.

He ran to the Tropical Nursery and burst through the doors.

"Where is the Spanish guy? Where is he?" he asked.

When he found me, he slightly altered his question:

"Where are they? Where are they?"

He wanted, *needed,* to see if what he had heard was true.

It was.

He was so excited, wearing the broadest grin I had ever seen, from ear to ear and beyond. I thought he was going to explode.

I hadn't realized how many seeds I'd gone through in my experiments. I was later shocked to find that the only known wild population had been destroyed; worse still, a rat had got into the greenhouse in Bonn and eaten the last plant.

At the time of my tortellini moment, I had been playing with the last five seedlings on the planet.

11

Hot Property

In 2010, the annual International Day for Biological Biodiversity fell on May 22. That was the day Kew chose to tell the world the *Nymphaea thermarum* story. It's one of more than a hundred plant species on the edge of extinction that survive only in botanical gardens around the world.

The story kept us busy for nearly two weeks. When you are being interviewed by almost every major newspaper and TV networks such as CNN, Al Jazeera, and the BBC, and talking positively about plants, explaining the problems facing biodiversity and creating awareness of plant extinctions, you know it must be making some kind of impact. *Nymphaeum thermarum* was suddenly a "pop star" plant, elevated to the status of an endangered exotic orchid, a plant that many gardeners would do anything to own.

People kept asking Kew if they could have one, but *Nymphaea thermarum* could go only to botanic gardens, not private collectors. All we could do was respond with a firm "no."

But when someone says "You can't have it," people want it even more. They grow obsessed. I think most gardeners are the same. It is about rarity, having something that no one else can have, whatever it looks like. If the Himalayan blue poppy (*Meconopsis betonicifolia*) was a roadside weed, we wouldn't want it in the garden because it would remind us of roadsides. It is the same for the daisies (*Bellis perennis*) that grow in the lawn. If it was difficult to get them to carpet your lawn, there would be tutorials on YouTube about how to cultivate them instead of the number of ways they can be eliminated. It is as much about wanting that piece of rarity as it is about the status that you gain from growing it.

People were even approaching me to ask about it. "Seeing as you propagated the plant and look after it at Kew," they wondered, "are you entitled to keep one of your own at home?"

"No," I'd say, "I'm not, so you can't have one either."

Then they would start complaining: "Why isn't this plant on display in the public greenhouses?" We reluctantly put twenty-two plants in one of the ponds in the Princess of Wales Conservatory, in an awkward, inaccessible place. By this point, it was widely known that it was extinct in the wild.

The only way to get one was to steal it. A few months later, one had gone.

It went missing on Thursday, January 9, 2014. Nick Johnson, the greenhouse manager at the time, was the first to notice the hole; after double-checking with the last stock check, it was obvious that one was missing. The raid must have been planned. The plant was down near the water's edge. To grab it, someone would have had to crawl along a railway sleeper, fight their way through some *Anthurium* foliage, and then hang over the mud. As Sam Knight later wrote in *The Guardian,* the thief must have been agile, and pretty brash, to scramble down there in front of the other visitors. And the thief knew what he or she was doing too: the plants were not flowering, so they looked like small, dull lettuce leaves. One of our gardening colleagues, who had been working all morning in a nearby pond, saw nothing.

But the thief hadn't exactly been a contemporary Pink Panther. The hole that he (or she) left was obviously a hand scoop: you could still see the finger grooves the thief had left in the mud.

About a month earlier, Nick had come across a young French visitor who had "strayed" among the plants, supposedly to take photographs. His backpack was full of plants. Most of them were obtainable from ordinary nurseries, but there was also a rare *Myrmecodia*—a Southeast Asian plant that forms symbiotic relationships with ants—that Nick had grown in Kew's conservation nursery. Nick reprimanded him and asked what he intended to do with the plants.

"Root them and sell them on the Internet," he replied.

The man just couldn't care less.

Nick's next call was to Kew's own police force. (Yes, Kew has a specialist force responsible for on-site security. Established in 1845, the Kew Gardens Constabulary was originally staffed by part-time gardeners and veterans of the Crimean War.) However, they don't tend to arrest such first-time opportunist thieves, so the officer took a photograph of the French guy, marched him to the exit, and told him to leave Kew and never come back.

Over Christmas, Nick, who was pretty upset by the whole thing, went online to search for sales of rare plants from suspect sources. He messaged a seller on eBay in California who was auctioning seeds of the critically endangered St. Helena ebony, which in no way should have been on the market. The aim of places like Kew is to ensure that if rare species are sold, the profits are shared with their country of origin to finance habitat restoration. Johnson tried to explain this to the seller, and pointed out that there were only two St. Helena ebonies left in the wild.

"Screw you," the seller emailed back. "This is capitalism."

When I *did* finally share the waterlily with a trusted grower in Thailand—under a contract banning its commercial use—someone photographed the plant with a long lens and the picture turned up on the website of a local nursery, offering *Nymphaea thermarum* for sale, even though buyers would receive a waterlily hybrid instead of the real deal. So even stealing a picture was the basis for a scam.

Nymphaea thermarum was internationally important, so the theft had to be officially recorded. The supervisor called the police, so there was a crime number for ease of traceability, but we didn't really want to take things further. If the stolen plant was to be sold, the authorities needed to know where it had come from. A couple of police officers came to the conservatory and took a statement, and a forensics team put on their white suits and crawled around the flower beds with magnifying glasses. All they found was the hair of a mouse in a crack in one of the railway sleepers at the edge of the bed where they had been planted. Among Kew's leadership teams there was talk of the need to install CCTV around the gardens, the kind of Big Brother approach that Kew had resisted for years.

❊❊ ❊❊

I had actually been away on holiday during the theft and flew back to the UK on the following Sunday. At eight o'clock on Monday morning, just before I came into Kew, the Metropolitan Police sent out a tweet proclaiming that the "world's rarest waterlily is stolen from Kew." That brief statement had all the ingredients of an Agatha Christie novel, and it inevitably led people to conclude that there was only *one* specimen of this rare waterlily left, and it was now in the hands of shadowy plant abductors. It was bound to capture everybody's imagination. And it did.

As I strolled into the nursery, unaware of the news, a telephone was ringing and ringing. One of Kew's press officers was on the line. "Everyone wants to speak to you, and there's someone from a gardening magazine waiting here now." My mobile phone rang—it was a journalist I knew. I put him on hold. The press officer continued, "Oh, you don't know? Someone stole a *Nymphaeum thermarum* from the display, and the Metropolitan Police have just tweeted about it." I realized an avalanche was about to come crashing down.

The world's press took up the story with gusto—the crime was hyped up like an art heist. The *Nymphaea thermarum* inevitably became known as "priceless" and the BBC's *Crimewatch* put out an appeal for witnesses.

Many good things came from it. The theft generated discussion. Subjects like the trade in endangered species and biodiversity were talked about in the newspapers for the first time in years. I spoke to the media nonstop for a week—I even woke up at 4 a.m. one day to talk to American television. In a strange way we benefited from the theft: we still had many other specimens of this plant, so that was not a problem, and the publicity we gained for our work was a bonus.

I am not surprised that someone stole the waterlily. What surprised me is that it took so long. But I do often wonder who it was and what his or her motives were. It certainly has commercial potential. It is small, doesn't need much water, and is the ideal hassle-free waterlily. As a houseplant, it would be pretty cool, and relatively simple to grow. Before the theft I was approached by a British nursery that said it would be willing to place an initial order for 25,000 of the plants from Kew, at $6.50 each. That is serious money. Sales would almost certainly go global—and it would probably do very well in Japan, where gardeners like miniatures. So if you were to scale it up and sell 2 million plants, they could gross $13 million.

Maybe it was stolen by a plant "geek." The Royal Botanic Gardens at Kew are world-famous, after all. The thief could have come from anywhere. There is no way of telling, but I think it was either an amateur, driven by the same obsession that keeps us all growing, or someone stealing on behalf of a rich collector, like art theft. It does happen in the plant world. I can understand why someone would do anything to get a plant; I know how it feels to want a plant desperately—indeed, my mother might say I "stole" my first waterlily! I see both sides: the passion of someone who is obsessed, driven by an addict's craving, but also the hurt of someone who has suffered the theft of something they care about deep in their soul.

It would be easy for a thief to justify his actions to himself. "There were twenty-two plants and I stole only one," he or she might say. That is one of the reasons why the debate was so interesting. After all, who do plants belong to? Didn't plant collectors harvest plants from the wild to furnish European gardens? Sir Joseph Dalton Hooker, one of Kew's most celebrated directors, André Michaux, who introduced 5,000 trees to France, and Robert Fortune, who brought tea out of China, all spent years scouring the world, plundering tens of thousands of plants—and made fortunes for those who sold them.

Botanical gardens were at it too, from the British at Kew to the Spanish at the Acclimatization Gardens in Puerto de la Cruz in Tenerife, the latter crammed with plants they were sending back to the Botanic Gardens in Madrid: the botanical booty of empire. The Dutch tulip industry was kick-started by the theft of bulbs from the gardens of Carolus Clusius, a botanist in Leiden, in the 1590s. In 1876, Kew paid £700 to Henry Wickham for thousands of rubber seeds that he'd smuggled out of the Amazonian rainforest. In Brazil, Wickham was known as the *príncipe dos ladrões* (prince of thieves) and the *carrasco do amazonas* (executioner of the Amazon). But in 1920 he was knighted by George V.

Not everyone was sympathetic to our loss. I read some of the responses to Sam Knight's excellent article for *The Guardian*. There was the cynical: "A plant thief is just somebody who doesn't work at Kew." There was the dismissive: "Their stories tend to be glorified." And there was also, of course, the ill-informed generalization: "Personally, I have no worries about what has happened, I feel there is an arrogance about Kew. They deserved what they got." Someone even came up with the conspiracy theory that I stole it myself to create publicity for what I was doing.

It's difficult to assess how much plant theft really matters—habitat destruction is a much greater threat, and there are fewer laws to prevent it. I read that Dr. Henry Oakeley, who was once president of the Orchid Society of Great Britain, had visited the remains of a forest in the Peruvian Andes, one of the last known locations of a rare species of *Anguloa* orchid. All that was left was an area of 100 feet by 300 feet, and it was being slashed down, piece by piece, to grow maize. He would have been put into jail if he had tried to collect any plants and save them, but the farmer was perfectly entitled to hack it down.

We have now sent some of our *Nymphaea thermarum* plants to other gardens, and perhaps in time this waterlily will enter commerce, not because it was stolen, like the bottle palm from Mauritius, but through the correct channels so that those who deserve to benefit will do so. About 20 percent of the world's 380,000 plant species are now thought to be threatened by extinction, and as more plants become endangered, their rarity makes them more and more desirable as "trophies" among collectors.

Richmond police closed their investigation within a few weeks. I suppose drugs and terrorism are a higher priority.

⋙ ⋘

You may well ask: What is the point of saving a species that lives in a marginal, tiny habitat that no longer exists? Or as Louise Mensch, the former Conservative MP, put in a tweet: "Got to say what's the point? Ordinary looking plant hardly worth saving. #Darwin."

No one even knew that *Nymphaea thermarum* existed until the 1980s, and when I started working on it I had no idea why it was so important. Yet once it was cultivated at Kew, we realized something quite remarkable about it that could not be detected from its appearance.

When researching biology and related sciences, you often need a "model." This means that everyone uses the same species to conduct experiments—it validates the results. Rats are used in medicine testing, fruit flies in animal genetics, and thale cress (*Arabidopsis thaliana*) is the standard genetic model for plants. While *Arabidopsis* is very useful—it was the first plant to have all its genes decoded; it goes from seed to fruiting in less than a month, so you can have many generations in a year; and its size means that you can grow lots of them in a small space—it also dies almost immediately after flowering.

If only there was something that lived longer, that would open up opportunities for long-term experiments like genetic alteration . . . It had long been felt that another plant was needed, one that was lower down the evolutionary "trunk" of flowering plants. For years it had been impossible to find a species to fit these requirements.

Then they found *Nymphaea thermarum*.

After running extensive molecular tests, it turns out that

N. thermarum not only has a relatively small genome (the genetic material of an organism), but you can grow one hundred plants in a square yard, it reaches flowering in two to three months, *and* it can live for decades, making long-term experimentation possible.

Now *Nymphaea thermarum* is being transformed into a dream genetic model, with a dedicated team of researchers working around the world on it. Sometimes the possibilities are just as invisible as the CO_2 that makes them grow.

Darwin would have been thrilled.

12

Bolivian Botany

I am always in standby mode. Conservationists can never relax. You have to be ready for that call to fly out to anywhere in the world to save a plant, or to help teach people how to take care of it. After several months of waiting, we were suddenly booked on a trip to the Pando area in Bolivia, an isolated part of the Amazonian region close to the Brazilian border.

The last few days before leaving are always frenetic. I have to organize everything that is needed for my trip and plan for my absence at Kew. I have to make special arrangements for the care of plants that may struggle while I'm away, taking cuttings as a precaution against worst-case scenarios and giving detailed instructions. It's a marathon before I even get on the plane.

This Kew project was being led by the Americas and Natural Capital teams of the Science Department, who were researching the flora of South America; it was sponsored by the smoothie company Innocent. One of our aims was to teach the local people a system known as "alley cropping," using a tree called *Inga edulis* to regenerate the areas that had been badly damaged by logging or abandoned after grazing by cattle. The idea was to increase the range of crops they could grow, which would improve their diet, give them crops to sell, and reduce their dependence on plants in the wild.

Inga is a group of tough trees in the pea family. The pulp in the seed pods is sweet and rich in minerals—it tastes a bit like cotton candy with a hint of banana and can be eaten straight from the pod. The tree grows super-fast and draws nitrogen from the air by an association with bacteria in its roots. In most agricultural land, or in your garden at home, nitrogen is stored in the soil, and if levels are low we add manure or fertilizers to correct them. In the Amazon it is different. The nitrogen is stored in the plants and leaf litter, so if you hack them down and burn the rainforest, the few nutrients and minerals in the soil get converted to ash and are rapidly washed away by the heavy rain. Before long, little grows, and because it will not support grazing cattle, the land becomes redundant.

Inga trees help to make the land productive again; if conditions are ideal, they grow incredibly fast, and they flower twice annually. When planted in rows they provide shade and improve the ground with their falling leaves, which rot and increase the soil fertility, so crops like sweet corn can be grown between them in the rainy season. Eventually the ground is suitable for farming or cropping, and when the trees are pruned they produce firewood. Planting them is a simple and effective regeneration method and has been successfully used throughout the world. Yet again, our survival is dependent on plants.

After Kew has built the nurseries, they need someone to teach the local people how to run them. That's where I come in. My job is to go from place to place, spending a day with each group, showing them how to make the most of the nursery and, more specifically, how to propagate plants, using techniques like taking cuttings, which they often haven't heard of.

I still find it surprising that although the farmers live in and are from the forest, they don't have much clue about how to grow plants. They seem to know how plants can be used, where the best plants are found in the jungle, and when to harvest them, but not how to propagate and grow them as crops. That kind of knowledge would change their lives. Some of this is explained by the fact that some of the people living there originally came from the Andes, where they grew crops more suited to the Andean climate, like tomatoes and potatoes. As recent arrivals, they don't have the inherited knowledge to farm the local area. Others are descendants of indigenous people who were mainly hunter-gatherers, collecting everything they needed from the forest around where they lived. The rainforest is their larder, chemist's shop, and builder's merchant, yet it is still being chopped down.

Before we could go about helping the Bolivian villagers, however, we had to actually get there.

The journey from London to La Paz in Bolivia en route to Pando involves several plane journeys, thirty-six hours of sleep deprivation, and one sledgehammer blow of altitude sickness. At 13,000 feet above sea level, La Paz is the world's highest international airport. I thought I was hallucinating when we were met by "cholitas," women from the Aymara and Quechua communities who wear bowler hats and more layers than an onion.

The sense of the surreal was compounded when I saw an ad in the airport shop proclaiming, "Coca, the healthiest thing for you"—quite a different opinion from that of doctors in the UK, but people throughout the Andes chew the leaves of the coca plant and have done so for millennia. It is part of their culture and essential for survival at altitude, where the lack of oxygen can cause light-headedness, drowsiness, and, in the worst cases, death. Wherever you turn there are people, cheeks swollen like hamsters, chewing what looks like half a kilo of coca in each side.

The coca plant may look rather boring, like a cross between a blackthorn bush and privet, yet it is anything but. When you hold a leaf against a strong light, you can see that, along with veins like any other leaf, it has two especially long ones, one on each side, parallel to the edges—a unique characteristic. The small flowers produce lots of tiny red fruits, giving it a rather Christmassy look, but the "white stuff" that is made from the leaves has caused more mayhem and disruption to societies around the world than any damage done by real snow.

There are two species in use, *Erythroxylum coca* and *Erythroxylum novogranatense,* both with two varieties. DNA analysis shows they are all closely related and were domesticated long ago, in pre-Columbian times. Wild populations of one variety of *Erythroxylum coca* have been found in the eastern Andes, but the other three varieties are known only as cultivated plants.

Some facts about its artificial selection are surprising, to say the least. One relatively new strain, *Boliviana negra,* known as *supercoca* or *la millionaria* (the millionaire), is resistant to the herbicide glyphosate, a key part of the government's multibillion-dollar aerial eradication campaign known as Plan Colombia. So far the herbicide resistance has two possible explanations: one is that a network of coca farmers used selective breeding to enhance this trait, through trial and error. *"Paciencia es la madre de la ciencia"* ("patience is science's mother"). The other theory is that the plant was genetically modified in a laboratory. In 1996 a patented glyphosate-resistant soybean was marketed by Monsanto, so it is quite possible that coca could have been genetically modified in a lab too. It seems growers will always stay one step ahead of the authorities. As they say, *"Poderoso caballero es don dinero"* ("money is a powerful lord").

According to research, coca leaf, in its natural form, does not induce a physiological or psychological dependence. It is worth remembering too that the original recipe for Coca-Cola contained extracts of coca. Taken in the traditional way, it does not seem to cause problems.

Whatever your view, we can agree on one thing: never underestimate the power of a boring-looking plant.

❧※❧

Despite the shock to our systems from the altitude, we managed to make it to Pando without needing any coca leaves. Soon enough, we were ready to start the program; we got into the jeep and began to make our way to the first village. I was the only thing hindering our progress.

We found a waterlily almost as soon as we left Pando. Barely an hour from the town, we crossed a slow-moving river where a large pond had formed nearby.

Sometimes I feel like a waterlily-hunting robot. Whenever I detect water, I can't help but do a laserlike scan to check what aquatic plants might be there. I can spot waterlilies instantly. "Waterlilies! Can we stop?" is a classic Carlos line.

The group was desperate to get to our first destination. We were already late. After my shout, the jeep sped on. I looked back in the direction of the pond, then at the driver, my face etched in a mixture of angst and anger.

"You can't do this to me!" I said.

They had to understand that I am a waterlily addict and I needed my fix. Denial means I suffer withdrawal symptoms: frustrations, tantrums, and pain. A few feet later, the vehicle lurched to a halt and the driver said: "Okay then, but as quickly as possible."

I jumped from the car, sprinted toward the bridge close to the road, and slid down a slope to the edge of the pond. There was a lady washing her clothes, right in the middle of where most of the waterlilies were. She looked on in amazement as I took off my walking boots—a declaration of intent—and waded out into the pond, picked a waterlily, then ran back past her to the jeep before we sped away.

Even though the flower was closed, everyone in the car was shocked by the pungent smell of acetone it was emitting. It's the kind of smell that makes your eyes itchy and tickles your nostrils.

I looked at the plant. There was a ring of hairs at the base of the petiole—the stalk that connects a leaf or pad to the stem. I didn't make much of it at first; I knew that there was one species with that defining trait, but I couldn't remember the name. All I knew was that this species was a subgenus of *Nymphaea* called *Hydrocallis,* the largest subgenus of waterlilies on the planet. It contains about twenty species, which are found in the wild in the Caribbean and Central and South America. They are night-time party plants that flower at crazy hours, some between 1 a.m. and 5 a.m.; others flower on the first day for a couple of hours, typically between 8 p.m. and 10 p.m., then on the second day they open from 8 p.m. till dawn. Their classification is complex and there are not many in cultivation. I have seen more pressed flat as herbarium specimens than living plants, though a wilted dead plant won't help you much when you actually see a living plant.

Knife in hand, car moving, I sliced the bloom in half while answering questions about the scent. What a surprise I got. This group of South American night bloomers is distinguished by its carpellary appendages (which make a small cage over the stigma, or female parts). The classification of the plant is based on the shape of these appendages—they can be cylindrical, conical, or tapered, and of different lengths and colors. Normally they are white, though in some species they are pink or red. These were purple. I couldn't remember ever reading about purple appendages. The stigma was bright yellow. I didn't think I'd seen this type of waterlily before—it had to be a rare species (rare in cultivation, rare in books, and maybe rare in the wild too).

A good two hours later, we stopped for lunch. Agouti (a type of South American burrowing rodent) or fish were the dishes of the day. I went for fish but also asked for an old newspaper and, with that, I pressed the waterlily.

Two days later, I had an epiphany: could it be *Nymphaea amazonum*? This was a common species, but I could not recall seeing any with deep purple appendages. On the return journey to Pando, several days later, we passed the pond. It was pitch dark—jungle dark.

"I know it sounds weird and we are all tired, but I really need to stop again," I said.

"But it's late," came the reply.

I promised to be quick, and scrambled down to the pond. This time there was a flower that had just started opening, but there was not a seed to be found. The lady I'd seen previously had been replaced by a group of party-goers looking at me from the bridge; I wonder if they thought I had downed even more drinks than they had.

When I collected the first flower, there had been only one bloom among the thirty or forty plants. It was a second-day, male-stage flower, and, just like today, there had been no females in the pond or the section of river nearby. So perhaps there were not enough flowers to achieve pollination and if, occasionally, there were, there were perhaps not enough to attract the number of pollinators needed to ensure success.

This was important information. Some species, such as *Nymphaea amazonum*, are self-pollinating and produce seeds freely. Lack of seed could perhaps point in the direction of a species that doesn't self-pollinate. But there were further possibilities. It could be a naturally occurring hybrid, and could be sterile. Or, if I was lucky, it could be a new species. Back at the hotel in Pando I did some research. My collection did not match *N. amazonum,* and none of the species I looked up mentioned purple appendages. Since we had permits to collect, but not to export, I gave a couple of the small plants I had collected, together with some herbarium specimens, to the Santa Cruz Botanic Garden, who shared the plants with a waterlily enthusiast that we both knew, so that they could investigate further. We needed to find out if the plant produced seed after cross-pollination. If it didn't, it would be a hybrid for sure; if it did, then it was likely to be a new species.

Later I managed to confirm that there was no other plant with purple appendages. *Nymphaea novogranatensis,* in Venezuela, has deep crimson appendages, and *Nymphaea lasiophylla,* in eastern Brazil, has appendages that are reddish-purple at the tips, but nothing matched completely the specimens I found. I finally concluded that my find had to be either a hybrid or a new species. Bolivia has the greatest diversity of waterlilies in the whole of South America, with about ten species, so it isn't such a crazy idea that some are still waiting to be discovered.

I learned how close Pando was to the Brazilian border on the first afternoon we arrived. I went for a walk around the town and wandered across a fancy bridge. It was not until I noticed that people were speaking Portuguese that I realized I had strolled into Brazil. Luckily there is a neutral zone a few miles either side of the bridge before you need to report your presence to the local authorities.

The first day—the day I found the waterlily—we were driving to a remote village called Palacios, by a lake of the same name. When we finally arrived there was hardly anyone there—a local soccer game was more of a crowd draw. I took the chance to visit the nursery while it was quiet, and a local showed me around, pointing out a newly established *Inga* plot that still needed constant weeding as the trees had not grown large enough to shade the ground. There was a cropping alley (a row of widely spaced trees, with a companion crop growing in between) several miles long that went straight to the village. Its position was beautiful. Part of it was set near the U-shaped oxbow lake, and the local told me that there were giant water-lilies in there. I couldn't see any, and the water seemed a little low, but he insisted the waterlilies were there. Ten minutes later I was looking at a couple of small plants that were large enough to flower. It was *Victoria amazonica.*

The annual rains always bring floods, and you could see head-height watermarks on many trees. I asked how the people coped, and the answer was simple: "We change our cars for boats." The road to the village becomes a channel, almost like a tropical Venice, and they go shopping by boat. I guess it breaks up the routine, though on the downside the rains raise the mosquito population to an unbearable level.

We left the village after a brief stop and headed off to the next village on our itinerary, Motacusal, where we would be able to see some Brazil nut trees.

We'd been traveling for a couple of hours when we suddenly swung across a field to reach a major new road being created by machinery ripping through the rainforest. One machine was knocking over giant trees, those majestic kings of the forest, as if they were sweet corn plants. As they fell, other machines were following behind, compacting the soil with giant rollers. The pace of destruction was horrific.

At least Bolivian law forbids damaging Brazil nut trees—

Brazil nuts are a major part of the local communities' income, and you can't cut down one of these trees without permission. Builders will divert the path of a road rather than cut one down, so it's not unusual to see a magnificent tree standing defiantly in the middle of a highway. Despite it being called the "Brazil" nut, Bolivia is the biggest exporter of these nuts, accounting for about half the world's annual supply of about 20,000 tons of nuts. The tree is one of those that holds up the "ceiling" of the forest, and is often the tallest. The wood is good quality too, but it's the economic value of the nuts that led to its protection by law.

Locals can make up to $150 a day during the harvesting season—often a family's main income. Every nut is harvested from the forest—they are like money growing on trees. Locals call the tree *castaño*—the very same name we use in Spain for sweet chestnut, whose Latin name is *Castanea sativa.* Harvesting nuts from the wild is one of the few great examples of forest exploitation that does not involve logging.

You may wonder why the Brazil nut tree (*Bertholletia excelsa*) has never been cultivated commercially outside its native habitats. The key reason is that the flowers of the tree, which open for just a single day each year, have bulky petals that can be pollinated only by a bee that is big enough to push through them and reach the nectar inside. *Eulaema meriana* bees (also called orchid bees), mostly the females, are the only bees big enough to do this, and they are not found outside of these same native habitats. The males of this species have a different role to play. They harvest the fragrance of some orchids, including the genus *Coryanthes,* to increase their chances of mating with a female, and these orchids need preserved areas of old forest in which to grow. It is an extremely intricate ecosystem.

In a further twist, agouti—the very same large rodent that I was previously offered for lunch—are the only animals capable of chewing through the hard, woody, outer coat of the husk to get to the seeds (i.e., the nuts), though capuchin monkeys have been seen throwing them at rocks until the fruit breaks open; humans use a machete. The agouti buries seeds around the forest for later, just like squirrels in Europe and North America. In this way, they disperse and plant the seeds, making them probably one of the most valuable rodents to humankind.

Brazil nut tree plantations have largely failed, and growing Brazil nut trees away from their natural habitats is not possible because there would be no pollinators. The areas of forest I visited were a mosaic of degraded sections, patches of secondary forest—forest that had been previously logged but had now regrown—and some small but fairly pristine patches. If the degraded areas and forested areas with a low density of Brazil nut trees were replanted and restored using *Inga* forestry techniques, then planted with Brazil nuts, the pollinators would move from the good patches of forest where they still live and pollinate the Brazil nut. It would restore the forest and its ecosystems and provide an extra income for the local people, who would then conserve the parcels of forest because of the value they produced.

I wondered why this was not being done. Before I could ask, though, someone at Motacusal had asked an even bigger question: "Please can you help us to grow *castaño*?" It turned out that the seeds of the Brazil nut, which can take more than a year to germinate, often fail to do so. They are worth a lot of money to the people, and low germination means lots of wasted seed and disappointment.

It was time for my first training session at Motacusal, and as I started to explain how to take cuttings from other plants, the locals immediately wanted to know if this would work for the Brazil nut tree too. I had no idea. But I knew that many of the members of this plant family could easily be propagated that way. I thought hard before saying, "I think so." That didn't sound very confident, so I added: "Let's try."

Because Brazil nut trees are generally 150 feet tall and lack lower branches, a human ladder—like the one we tried in Mauritius—could not be safely attempted. Luckily, one of the local project managers had anticipated the problem and located a younger tree. Instead of climbing, we managed to harvest enough cutting material from this younger sapling. When cuttings are exposed to heat, even when bagged up wet to prevent desiccation, they very quickly become unusable. They were going to be bagged up for several days while we distributed them to the nurseries. If they failed, I failed too, and the locals would think of me as nothing more than a European spin doctor.

We returned to camp just as it was getting dark and the Bolivian cicada orchestra started played their psychedelic nightly symphonies. A feast was ready. We were led to a kitchen with a dirt floor and ducks and chickens wandering around. There was no electricity, so I ate my beef and rice in the soft glow of candlelight and planned how I could get the locals to think like a messiah.

❧ ❦

The next day, in a forest clearing in the village of Motacusal, I gave a talk about propagation. The blackboard was a piece of paper stuck to a tree with tape—I had to walk around the buttress roots of the tree to get from one side to the other.

I could see the locals thinking, "Here is this gringo—what is he going to tell us, then?"

The ice was broken when one of the locals turned up with a pair of domesticated coatis. They are racoonlike animals with pointed faces and long tails that seem to get everywhere. They climbed all over the people in the audience as they listened, and upturned every piece of equipment on my table: secateurs, plastic bags, even my propagation material. The audience livened up and by the end they would not stop firing questions at me.

The innocent-sponsored nurseries are built from a combination of local and imported materials. Small plants are shaded by frames covered with palm leaves, and the plastic water tanks and watering cans are often stolen. Lots of plants can be grown with minimal resources once you know what to do, though. You just have to improvise. Before leaving, I went to a roadside warehouse and bought some transparent plastic and rolls of the thickest wire to make a moist environment for the cuttings. The locals had told me that most of their cuttings failed, whatever the species. I was hoping that plastic, for once, instead of littering all the areas around the village, would help them to grow whatever they wanted to.

It was time to put my plan into practice. All I needed to make the framework was a machete, a couple of small posts, a thinner central cane, and some thick wire. Then I would drape plastic over the frame, tie the ends of the plastic like a sausage with string and weigh it down with a stone, add a bit of water, and the humidity-saturated shaded rooting space would be done. I did this, but unfortunately, by this time, all my Brazil nut cuttings had wilted. We put them in the propagator though, and they started to recover by the next morning—something that delighted the locals. I knew I would be very disappointed if this didn't work. It turned out that my shaded rooting space was the least of our concerns, however.

On our journey to the next village, Monte Sinai—which, from the map, looked as though it was at the other end of the universe—the car broke down. We were rushing to catch a ferry across the Madre de Dios River (Mother of God), about four or five hours' drive away, and had to wait almost a whole day for a taxi.

For me to be stuck in the rainforest with a bunch of Kew botanists wasn't too much of a hardship. We spent the day botanizing, challenging one another to identify plants.

"What species is it?"

"You don't know?"

"Okay, so what family is it?"

We came across one plant with three leaves that had been partially eaten by caterpillars and still managed to work out the family and genus. The biodiversity is so great in the area that it was impossible to name everything, other than their family or genus.

By the time the replacement four-wheel-drive arrived, we were six or seven hours behind schedule. We pressed on to get to the crossing for the Madre de Dios River. The last ferry had just left but, on seeing us pull up in the distance, the captain reversed back to the shore and let us board.

While I was at the pier, I'd noticed that all the ladies on the ferry seemed to be fidgeting vigorously and whipping themselves with cloths. I soon found out why.

Once you get on the flatbed open ferry, only the driver is allowed to stay in the vehicle. The minute you step outside, masses of sand flies take you on. Sand flies know the ferry is a rich source of food; however much you try to cover yourself, they seek out every millimeter of bare skin. Hands, face, ears— I got well and truly drilled. Nevertheless, it was awe-inspiring to see the size of what is described as a "small" tributary of the Amazon. The Atlantic was 2,500 or 3,000 miles away, and the river was already vast.

A major bridge was being built across the river by the Chinese, and the smell of gasoline hung in the air. Increasingly, not-so-safe petroleum wells are also being built. This, together with both legal and illegal gold ore extraction, spells bad news for nature in the area.

As we reached the middle of the river, the sand fly problem diminished, but just before we saw the back of the last fly from one side of the river, we started getting attacked by flies from the other. Docking at the opposite side, we were confronted by a steep slope and, with no space for the jeep to gain momentum, we were forced to walk up the path to meet the car at the top. Once there, we all jumped in, cranked up the engine to top speed, and accelerated away.

When we finally reached Monte Sinai it was early evening. We suggested the training session should be postponed, but one of the locals said, "Cancel it? No way! I am going to the next village to get a pound of coca leaves, then everybody will be happy for you to do the training now."

I didn't want to refuse, but after almost three days of solid traveling I was spent. We hadn't pitched our tents, had had nothing to eat, were covered in red dirt, and had been eaten alive by sand flies. I explained to the local organizer that I desperately needed some food.

"My wife will cook you beef and rice," he said.

"Oh," I thought, "not again!"

In the end we had to go to a shop anyway, where I tried to convince everyone that a switch to chicken just for one day wouldn't kill us. To my relief, many agreed and we had roast chicken instead.

My next question was: "Is there anywhere to have a shower?"

"Of course," the local organizer said. "Come with me."

I grabbed my kit and walked off into the dark rainforest. The "shower" was a few barrels of water. There I was, naked in the dark, sloshing water over myself from a bucket. When I finished, I reached over, put my hand into my bag, and pulled out . . . a nylon mosquito net. My towel and the net were in identical bags, and I'd picked up the wrong one. They are not the easiest things to dry yourself with.

I shook myself like a dog, managed to get into my shorts and flip-flops, then staggered back in the direction of the towel. Some villagers were standing around, and as I walked past them I heard them shout, "Carlos, Carlos, the training!" It was nine o'clock, and I was tired and disorientated.

"Come on, guys," I thought. "What happened to that ancient Spanish principle, *mañana, mañana*? Can't we apply this for once in my life?"

It appeared not. About twenty people came to my seminar—they sat in the flickering light, their bulging cheeks stuffed with coca leaves and their eyes wide open, with an expression that was slightly too enthusiastic. After demonstrating how to take cuttings, I asked for a brave man to come forward and have a go himself. The one who volunteered set to work as if it was a martial art, with jerking, robotic actions. Hack, chop, split. Practicals have never been so dangerous.

By question time they were frantically jabbering away at a thousand miles an hour, asking questions with urgency and talking loudly over one another. Before I knew it, it was midnight. I showed them how to make a propagation tent with wood, a narrow pipe, and a machete, and, sensing a chance to exit, concluded, "And you can have a go at all this tomorrow."

"No," they told me, "we want to do it now."

We had to go to the nursery, two miles away. In total there were about thirty people in two jeeps and on two or three motorbikes. Goodness knows how they managed to drive those bikes, given the state they were in.

Soon enough I had showed them how to make a propagation tent, *Blue Peter*–style—albeit using a machete. I was utterly exhausted as dawn lightened the sky.

"So, do you have any final questions?"

"Why don't we go back to the village, get some more coca leaves, sit here and watch them root?" one of the villagers said.

Tempting as their proposition was, we had a message to spread.

❊❊❊

Our next destination, the village of Remanzo, was a couple of hours' drive away, down muddy, unmarked tracks. This time our audience was a mixture of young and old. I knew it would take a while to gain their confidence. Once they understood that I had come to help, not to tell them what to do or sell them anything, they would relax.

I had picked up how obsessed the people of the Amazon were with citrus fruit, and this village was no exception. There we were, enjoying all the wonderful exotic fruit the tropics could provide, and all they wanted to do was grow lemons and tangerines. It was almost like the Mauritians, who were constantly asking me how grow tulips. One person's exotic is another's everyday.

I was amazed to find citrus plants growing in the Amazon. I wouldn't have been surprised to find them in the Andes, as it gets cool there at night, but here they were in the rainforest, flowering and fruiting too. Apparently they were introduced by the Spanish. I didn't have a gin and tonic, though. One of my favorite drinks while I was there was one they made from the white pulp that is found between the dark brown seeds of the cocoa pod.

I badly needed to freshen up. The bathing facilities were

little more than a pond, about 30 feet by 150 feet, surrounded by lush vegetation, with a bridge running across the center. I took off my clothes, put on my goggles, and dove in.

As I swam among the aquatic plants, I saw small, brightly colored fish that I recognized from my younger days when I had an aquarium. There were cichlids, different types of characins, and bottom-dwelling *Corydoras*. All of a sudden I noticed a large shadow, much bigger than a fish, and heard a noise. It was a girl on her way to school. She had stopped to see what was going on. It couldn't have been a common sight—a man wearing goggles and splashing around in the local pond, au naturel.

"Oh, hello," I said awkwardly.

"Hello," she said in return. She seemed less embarrassed than I was.

I wondered about the impact shower gel would have on the water quality of the pond. I figured the locals used soap too, and as the water was regularly replenished by rain, the concentration of soap would be diluted. As it was, the water was hard and did not make much foam, and the fish seemed mildly amused, nibbling my legs.

I was happily bathing, thinking, "For once I am in water, in the Amazon, and there are no crocodiles, piranha, or pacu to worry about" (pacu are a large, vegetarian relative of piranha, an important agent for dispersing seeds from riverside trees and one of the few fish that can carry seeds upriver, against the current; legend has it they have mistaken testicles for fruit). Surely nothing could hurt me here.

I was in the middle of rinsing my back when I remembered a BBC documentary. It said that it is very dangerous to swim in a small pond if there are any electric eels in it, because if they release their electrical charge, it will shock the whole pond. I tried to stay calm and finish my bath.

A couple of months after we returned to Kew, our partners in the Amazon revisited all the villages and sent us an email with pictures of the plants the locals had been growing.

Hi Alex and Carlos

All the cuttings are well—you can see the new growth. Everyone here has realized that your methods work and the communities are really happy with the new techniques. They've realized the potential to create many saplings of whatever fruit they want to produce.

Depending on the species, 60 to 100 percent of the cuttings were growing well, and in several photos roots were poking out of the bottom of the compost bags. A Brazil nut cutting had a nice rosette of fresh growth already popping up.

Not a bad result for a project based on a sheet of plastic, a knife, and a bit of local interest.

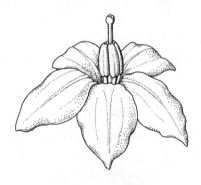

TOP: Proudly posing in front of my first pond, dug by my father at the *finca*, showing early signs of becoming a geeky-looking natural scientist.

BOTTOM: In my early days at Kew as a student, about thirty years old, focusing my obsession on *Ramosmania rodriguesi* and trying to pollinate the "living dead." *(Kew archives)*

TOP: The Temperate House at Kew showing its magnificence. At 628 feet long, it is the largest surviving Victorian greenhouse in the world, twice the size of the Palm House. *(Kew archives)*

BOTTOM: A view from the Black River Gorges National Park on Mauritius Island.

TOP: A flower of the Mauritian endemic species *Roussea simplex* being pollinated by the blue-tailed day gecko *Phelsuma cepediana*. (Dennis Hansen)

BOTTOM: *Trochetia boutoniana* is the national flower of Mauritius.

TOP: *Phelsuma* gecko pollinating the Mauritian endemic species *Trochetia blackburniana*. *(Dennis Hansen)*

BOTTOM: *Chassalia boryana*, or Bory's coral tree, is another highly endangered and beautiful Mauritian endemic, currently being cultivated at the Royal Botanic Gardens at Kew. *(Dennis Hansen)*

TOP: The last wild surviving specimen of *Ramosmania rodriguesi* in its triple concentric fence at Rodrigues Island.

BOTTOM: The endemic species from the Mascarene Islands *Terminalia bentzoe* ssp. *rodriguesensis* has heterophyllous foliage, meaning that different types of foliage occur on the same plant.

The last-known living specimen of *Hyophorbe amaricaulis*, which is grown at Curepipe Botanic Gardens on Mauritius Island, is often referred to as the "loneliest palm tree on the planet."

TOP: View of Rodrigues Island and its reef lagoon. Most terrestrial ecosystems of Rodrigues are highly degraded; however, the conservation efforts and restoration areas provide great hopes for the future of the island's ecosystems and the remaining surviving species that occur elsewhere in the world.

BOTTOM: The giant pads of Victoria are known for being able to support heavy loads. In this picture, a not-so-large (by Victoria standards) *Victoria cruziana* easily carrying my son, Matheo, at the age of eight months at Kew's Waterlily House.

TOP: A wild population of *Victoria amazonica* in an oxbow lake in Amazonian Peru.

BOTTOM: Here I am teaching the basis of plant propagation in the jungles of Pando, Amazonian Bolivia. In the picture are some of the villagers from Motacusal, who greatly rely on Brazil nuts as a primary source of income. *(Alexandre Monro)*

Bertholletia excelsa, the Brazil nut tree. As this tree is protected by law, it is often the only one left standing while the rest of the ecosystem is destroyed to log for timber, develop roads, or provide more space for cattle farming. *(Alexandre Monro)*

TOP: A *Puya raimondii* flowering stalk after being burned. The person on the right can give you an idea of the sheer size of the "Queen of the Andes."

BOTTOM: *Bomarea dulcis* climbing over the dwarfed *Polylepis* in a forest in Ayacucho, Peru, at an altitude of about 4,000 meters. The epithet *dulcis* means "sweet" in Latin and refers to the red, sweet-like seeds it produces.

TOP: *Cumulopuntia sphaerica* growing in the coastal, foggy environments of San Fernando, Ica, Peru. While the rest of the vegetation is dead or dormant, waiting for better conditions, this relative of the prickly pear blooms.

BOTTOM: Félix Quinteros posing next to *Prosopis pallida,* a magnificent specimen of a millenary huarango.
(Oliver Whaley)

TOP: *Neoraimondia arequipensis* is the world's largest multibranched shrubby cactus and is endemic to Peru. *(Oliver Whaley)*

BOTTOM: The fruits of *Nymphaea* burst and disperse their seeds when ripe. This generally happens early in the morning. *(Christian Ziegler)*

TOP: *Nymphaea thermarum* is the smallest known species of waterlily in the world. You could easily grow a blooming specimen in a teacup.
(RBG Kew)

BOTTOM: This rather beautiful form of *Nymphaea lukei* is found in Dog Chain Creek, Kimberley. This species was described only in 2011.
(Christian Ziegler)

TOP: Collecting herbarium specimens of *Nymphaea* in Lake Gladstone, western Australia. *(Christian Ziegler)*

BOTTOM: Before the introduction of the common honeybee, Australian waterlilies were pollinated by stingless bees (bee species in the genus *Tetragonula* and *Austroplebeia*). Nowadays, native bees have to compete with the introduced and much larger honeybees. *(Christian Ziegler)*

TOP: An example of the Kimberley endemic of the bicolored population of *Nymphaea lukei* at Charnley River Station. *(Christian Ziegler)*

BOTTOM: The fragrant blooms of *Nymphaea* are very attractive to their pollinators. Flowers close to the creek edge are sometimes partially nibbled by adventurous grasshoppers that take their chances and jump from the shore, as seen in some petals of this *N. lukei. (Christian Ziegler)*

While collecting in Kimberley, chasing ponds and creeks all day long is
the standard, but at sunset you just look for a suitable area to camp.
(Christian Ziegler)

13

Peruvian Plants

Coastal Peru is a hauntingly beguiling place. The site of one of the world's driest deserts, it is also home to mysterious and ancient ecosystems such as the dry forest on the Pacific coast of northern Peru. The problem is that prolonged sunshine and a regular supply of groundwater are also the perfect conditions for growing food—today, a rapidly expanding agricultural export industry has cottoned on to this area as a massive "greenhouse" for filling our supermarket fruit and vegetable racks during winter months. As so often, most of the deforestation happened during the nineteenth and twentieth centuries, with the industrialization of cotton, sugar, and wine. Surprisingly, deforestation today is less likely to be driven by agro-industry, which is more minded to protect and reforest, than by migrants who chop the trees for firewood or to sell as charcoal. Man versus nature, again. Whatever the source of the problem, though, when the cost is precious desert habitats, there is a dire need to teach agricultural sustainability, particularly how native vegetation can protect and sustain people by protecting and sustaining ecosystems.

Kew has been backing research, conservation, and the restoration of precious areas in coastal Peru, and has achieved some excellent results in tough conditions. A fifteen-strong team is working on native habitats and programs for sustainable agro-industry, supported by local volunteers and enthusiastic students.

In 2013, four of us from Kew flew to Lima before traveling to La Peña in Salas, north of Lima, to spend a week there. Agriculture in this region is due to expand massively with the construction of a new pipeline ("H2Olmos"), which will carry water under the Andes to the coast from the Amazon—with all of the implications that will bring. The huge challenge for the team was to create a protected area of 24,000 acres of degraded dry forest, then nurture it back to health, with the help of the local community.

The key to understanding anything about restoring a forest and its ecosystem is "monitoring"—initially setting a baseline to try to work out where you are going—seeing which species thrive and which are dying; which species disperse seed and which factors destroy seed. To be successful, you need to see both the small picture, on your hands and knees, and the big picture, with historical satellite images. So, in the scorching heat, and under leafless trees, we launched drones, collected plants, and set up permanent plots, even scanning them with ground-based LiDAR (Light Detection and Ranging, a remote sensing method that uses lasers to map areas). Most important, we listened to local people to understand their needs and hear about the environmental changes they had experienced—the most significant factor being El Niño: that switch from drought to deluge.

The Kew team developed a "seed ball" system to restore the native forest. The whole community invested time and love by gathering and processing seeds from trees like *Prosopis, Bursera graveolens,* and *Capparis scabrida.* They then hand-rolled seed balls in order to plant them just before the next major rains arrive, which could be up to ten years away. It was our first large-scale attempt to regenerate forest from the tiny areas that remained. It was fascinating to watch them make the seed balls. They would collect the seeds of five species that grew together naturally, press them into a pellet of alluvial clay, roll it around in their hands until the seeds were sealed, to protect them from being attacked by insects, then bake them in dry shade. They were then buried in the deforested ecosystem they had come from, where they will remain until the rains come.

If you are a plant, the more you grow, the more CO_2 you absorb. Rainforests are the fastest-growing of all ecosystems (the warmth and humidity work like a propagator) and are the densest too, but plenty of rain and humidity means plenty of decay, so a percentage of the CO_2 always returns to the atmosphere. The rate of growth is slower in dry forests because, as common sense suggests, conditions are less advantageous— organic matter decays more slowly due to the lack of water. To find out how much CO_2 is taken in or lost, you need to assess each type of forest individually. We also needed to know how fast each species grew, the soil type, and which plants were growing nearby. Most important, we needed to find out which species the locals relied on, and which they impacted most of all, and this required serious attention to detail. We randomly selected a representative plot and mapped every single tree and shrub, measuring their girths at a standard point as well as their height and width. We also collected specimens for verification in Kew's herbarium and registered every plant species growing in and around the plot, including the locally famous "*cojones del diablo*" (devil's balls, or *Luffa operculata*), with its unusual seed pods, and the parasitic tropical mistletoe (*Psittacanthus* sp.), with its bright red flowers. We tagged each tree with a label, and we left behind weather stations, some as small as a can of tuna, that would record climatic changes for many years to come.

With all this data we can learn when and how much carbon the trees uptake from the air. But, most important, we can learn how the trees react to climate change, the droughts, and the El Niño cycles. My job was to be the tree labeler, following the methods we use back in Kew's Living Collections Department (given that the 70,000 types of plants we grow at Kew are all labeled, you can be sure we know a bit about this).

With the help of locals living in small communities, we traveled off-road on motorbikes. Peruvians in the outback of northern Peru seem to use motorbikes for everything—shopping, herding cattle, and also, apparently, picking up foreign scientists and horticulturists. Journeys could be treacherous as we sped across cobbles, sand, and prickly vegetation—though health-and-safety training instinctively taught us to shout *"Mas despacio, por favor!"* ("A bit slower, please!") into the ears of our helmetless motorbike drivers—but eventually we got there. Inevitably there were a few incidents—scratches from thorns, a bit of dust inhalation, and the odd dog chasing after us, trying to bite our ankles—but no one incurred serious injury.

We set up four large monitored plots in a few days. There were two teams, and within each team two people did the measuring, another would take the notes, and another did the tagging. While we labored in the scorching sunshine, the locals who had transported us there relaxed in the shade of a large tree, bemused by the crazy scientists and debating such things as whether the cowpat in the plot came from their cows or from those of rival farmers in the next valley.

This dry forest is a bit thicker than a savannah. The trees are closer together, and most woody species—the trees and shrubs—are only 10 to 16 feet tall. Most are also deciduous, shedding their leaves in the dry season, then growing some more when the rains come. Low rainfall means slow decomposition, so dead trees hang around much longer than they do in wet climates. Annuals and dormant bulbs will also spring back to life whenever the rains come, in this rather unpredictable climate.

You'll find lots of cacti growing there, from the columnar, multibranched *Neoraimondia arequipensis,* which reaches 23 to 26 feet high, to the smaller *Haageocereus* species, as well as the *Melocactus peruvianus.* This latter is a curious-looking cactus with a football-sized green body and a projecting flowering structure on the top, a bit like a hat, that slowly grows up toward the sky. This projection, called a cephalium, is covered in small fuzzy spines and white hairs, has garlands of small, bright pink flowers; and is a kind of feeding tower for hummingbirds. This species is usually found growing on the tops of rocks, or some- times on steep slopes among earthquake-shattered rocks. There is no soil and no water—nothing else can grow there. A tiny crack, in the shadow of the "nursing mother" cactus, is sufficient to hold a seed in place, guard it from the sun, and provide it with some protection against the drying wind. That's all that is needed to start the cactus growing, so that later on, when it is more mature, it can withstand whatever harsh conditions the desert may throw at it.

Melocactus also has a unique growth pattern. When it is about the size of a soccer ball, the green body stops growing and all the energy is diverted into the cephalium. At that point the cactus dedicates all its efforts to blooming and fruiting relent- lessly. The fruits, like small pink chillies, ripen halfway inside the cephalium, waiting to be picked up and swallowed whole by birds and then defecated out, it is hoped far away from its parents. What's more, the body shape changes depending on where the plant is growing. Some are like upright rugby balls, others perfectly round, while some are melon-shaped or like deflated beach balls. The cephalium grows slowly and can also differ in shape; some are very long or curve upward, especially when the plant is growing on a slope. This gives each plant a character of its own.

The lowland bajada, or slopes, are dominated by lots of spiny, woody plants, including a type of caper with edible fruits called *Capparis scabrida,* and *Cordia lutea,* a flowering tree with beautiful, bright yellow flowers. The slopes are also filled with *Bursera graveolens* trees, whose broken twigs smell intoxicatingly of incense; it is like there is a fragrance oozing from the deep cracks in the dry rock. It is uplifting to see the pink necklaces of *Bougainvillea peruviana* scattered over this otherwise gray and brown scenery.

When we'd finished labeling and recording the plants in the plots that were to be monitored, the locals took us back to the jeep on the motorbikes and we headed south to the Ica area, where our partnership organizations and most of the staff that run the project are based. Here I would be providing training on propagation and nursery management to local teams. One of my greatest achievements there was to teach the local communities the art of pruning by showing them that cutting a tree carefully with a cheap saw from the local market, in precisely the right place, would not kill the trees they used for firewood but would make them produce stems again and again. They used to slash branches off with a machete, which made the trees liable to fungal attack. My colleague Oliver Whaley often jokes about that day, describing the training session as an almost biblical scene, with me roaming in the desert, pushing through the sheep, goats, and the hot, dry scrub, looking for trees that could be used for a practical demonstration while being followed by a trail of disciples. After Ica I would go to the Nazca and Inca areas and visit a few nurseries on the way, to teach them some things. The idea behind these nurseries is to restore deforested areas using native trees and plants, and for local people to learn how to cultivate a wider range of crops to improve their diet.

They also aim to involve children in conservation and educational activities, so that this knowledge is passed down to future generations.

On our way to Ica we stopped at a vegetable plot built by the local communities to preserve their heritage crops. A Peruvian Inca Orchid dog, a breed of hairless dog, was waiting for us at the door. They are usually totally bald, but this one had a fluffy white crest of hair on its head, giving him the appearance of Stripe in the first *Gremlins* movie.

There was also a small cuy farm. Peruvians eat cuy—or, as we know them, guinea pigs. It's a traditional food, and often a source of panic when tourists find out what is on their plate. Here they were also kept to fertilize the vegetable plot. They were caged in long raised runs and fed all the weeds and food waste; their waste would then be dug in as fertilizer.

I was shocked to see some crops in a different setting, like "achira" (*Canna indica* syn. *C. edulis*). Canna species and hybrids are common garden plants in Europe and the USA. We never think of them as crops, but they have been used in this way since ancient times by many communities. The rhizomes are a source of starch, for humans and livestock; the stems and foliage are used as animal fodder, the young shoots as a vegetable, and the seeds are added to tortillas.

On another part of the plot they were growing cotton, and the bolls (seed capsules) had split open to reveal the fibers within, in a host of different colors: mustard yellow, crimson, grayish-pink, and tones of terra-cotta. I was used to seeing the white commercial cottons and their wild, darkish-brown relatives. But from pre-Inca times, local cultures selected forms of cotton for their bright color. The most difficult thing is to get it to grow pure white, but Western industry long ago demanded it to make it easier for artificial dyes to be used (like blue for jeans). When this policy was implemented in Peru there was a risk that local varieties would cross-pollinate with the white varieties, so the growing of local varieties was banned. Luckily some people disobeyed and these varieties survived; they are important for the patterns and styles of traditional clothing.

We carried on, heading south, following the Pacific coast; the further south we got, the drier the environment felt. While it's rare to find fresh water in Nazca and other nearby areas of Peru, there is plenty around, just in forms we often ignore. The plains hardly ever experience rainfall, apart from El Niño flooding, which occurs about once a decade. Rain falls mostly in the Andes, and rivers run from there to the Pacific. Some of the rivers are permanent, others temporary, and they can continue for a long time underground. It's here, along the rivers, where most vegetation is found—both native, such as *Salix humboldtiana* (a species of willow), and, sadly, invasive introduced species such as the Spanish *Arundo donax* (a giant reed with canes up to four meters tall).

But the place where the water is most abundant is, in fact, the air.

A foggy desert sounds like a contradiction, but fog is common in Peru; in autumn and winter it's like a permanent cloud layer. It's so thick that Lima locals go up into the mountains, not down to the beach, in order to sunbathe.

❖

One tree we worked on with the locals really captured my imagination: *Prosopis limensis,* or the huarango, one of a genus of about forty-five closely related species in the pea family. *Prosopis* are mostly spiny trees and shrubs, and are found in several subtropical and tropical locations including the Americas, Africa, Western Asia, and South Asia, though the vast majority are found in South America.

The huarango appears to have the deepest root system of any tree species in the world, as it can reach down beyond seventy-five meters in search of water. It can live for more than 1,000 years and is often found where no other trees can grow; its heartwood is believed to be the second-hardest wood on the planet.

Several plant species are great at capturing water because mist or fog condenses on them. Huarango is one of the best. Not only does it suck up water from deep in the ground, it also condenses the fog on its leaves and stems at night, which trickles down to the roots.

Plus, it has another trick: just like the *Inga,* it captures nitrogen from the air through its leaves, using a special bacterium in its root nodules. Most plants absorb nitrogen through their roots from the soil, where it has been released by decaying organic material, such as fallen leaves, droppings, and carcasses. Water aids this process, but in a desert the lack of rainfall means that the nitrogen available in the soil is rarely at a level that can sustain plant growth. It is, however, abundant in the air—and out of reach for most plants. But not for the huarango.

It is easy to see why this type of tree is so successful at coping with harsh conditions. No wonder it is an invasive weed in Hawaii and tropical Australia, where it chokes complete valleys.

You might think it is invincible, but you would be mistaken. Huarango is threatened in its native Peru.

It is a vital element of the desert ecosystem, with many organisms depending almost totally on it, like the slender-billed finch (*Xenospingus concolor*), whose populations have declined alongside the huarango. This little bird has a clear preference for riverine forest; when the trees go, the bird disappears.

The tree is also a valuable asset to communities. It provides shade for cattle and locals (though good thick footwear is essential, as the thorns accumulate in the soil and perforate thin soles). The branches can be pruned and used as fodder in time of drought, and the wood is used in construction. It also produces plenty of nutritious pods; locals call them *huranga* (a feminization of the name of the tree), and they are sweet with a hint of vanilla and are excellent for feeding to cattle. They can also be ground into flour, which can be used for making bread, desserts, and even ice cream.

And yet the trees are being felled for a very mundane reason, as a conflict brews between traditional and modern demands. Toward the end of the 1950s, a change in government economic strategy led to the building of thousands of intensive chicken farms in Peru and the beginning of a new era—that of the chicken barbecue. Enter *pollo a la brasa*—meaning "chicken roasted on firewood"—for which huge amounts of huarango were felled. The high calorific qualities of the chicken made it a popular dish, and the demand for huarango wood as firewood massively increased, despite the wood being hard to cut down. I saw plenty of intensive chicken farms when I was there, mostly by the beach, where there was a supply of water; huarango tree stumps were a common sight too.

When I first arrived in Peru, I was taken with this delicious chicken dish. It was only a few days later that I realized that the real price for this delicacy was the huarango. Even the most ancient specimens, known in Spanish as *huarango milenario,* the "millennial trees," have been felled as a result of this demand, and there has been a dramatic reduction in tree cover. Part of Kew's project was to map the green areas where it was growing with a drone, and quantify year after year how much was being lost or gained.

The era of the chicken barbecue, however, was not the first time humans had deforested the plains of Nazca.

The ancient Nazca civilization is famous for creating complex line drawings in the desert that can be seen in their entirety only from the air. They were created between 500 BC and AD 500 and depict animals such as monkeys and whales as well as geometric figures several miles long. The ancient Nazca people also formed a sophisticated society, with complex irrigation systems for agriculture. Despite the apparent skills and expertise of the Nazca, modern researchers now say that these ancient peoples inadvertently contributed to their own demise by the mass removal of the huarango. Originally, the abundance of trees reduced the impact of El Niño floods, the roots binding the soil together to minimize erosion and helping to replenish reserves of underground water. As the number of trees diminished, and felling reached a tipping point, the soil was laid bare and felt the full force of El Niño. Floods swept away what remained of the vegetation, the land became a desert, and the Nazca civilization disappeared. A sobering lesson for us all.

For us at Kew, the desert plains of Nazca are extreme in almost every way. They experience intense heat (often above 86°F), intense drought, frequent earthquakes, and, of course, almost biblical floods nearly every decade. Pachamama, the South American goddess of nature, shouts loudly and often goes hungry. Perhaps you would be furious too if someone cut down your precious trees—including your 2,000-year-old treasures—just to roast a few chickens.

For us at Kew, the first step in counteracting some of the damage was to work out a way of involving local people in replanting projects, and to set up *viveros* (nurseries) to produce trees for the future. Once this was done, a horticulturist was needed to provide training to several community-based *viveros*, scattered over the Ica and Nazca plains. This is where I came in.

꘏꘏

Traveling through the Nazca plains is quite something. Some landscapes are surreal, and took my understanding of the words *dry* and *arid* to a whole new level. You can go for miles without seeing a single plant—there is nothing but bare rock and sand. The nakedness of the earth is broken only by humans.

The area is often shaken by earthquakes of great magnitude, and the devastation and loss of life can be significant. As a result the government has allowed people to resettle where they can, so if you find a piece of land that is not occupied, you can construct a house or shelter and register it as yours. Initially this sounds like a good idea and, of course, for many it has been vital for survival. But over the years it has evolved into a cottage industry, and can do more harm to the environment than good. The phenomenon whereby you arrive with a structure and claim your land is known locally as *invasiones.*

The definition of a structure is pretty simple. Anything that is enclosed, with four walls and a roof, can be registered as one. They tend to be rectangular, with a permanently open door in the front panel—a bit like a beach hut without the beach. The invasive Spanish reed *(Arundo donax)* is cut and woven into panels, called *esteras,* to make these structures, and they are sold everywhere, including gas stations. You can buy as many as you can fit on the top of your car, travel around, find a plot of land you like, and stake your claim. In recent years this has been done en masse, with land traffickers staking out hundreds of acres of empty desert at a time—usually under cover of night. They then sell the staked-out plots, with or without fake paperwork, to students and Andean migrants looking desperately to find a home.

In some highly populated areas, such as the outskirts of Lima, many thousands of people are living in slums constructed from *Arundo donax,* divided from the rich in one district by a three-meter concrete wall called "the Wall of Shame." Packed with families who often have no access to basics such as running water, the slums of Lima may be treeless, but they are definitely full of life—though it is a life of constant struggle.

In other more rural and remote areas, especially near the main roads, thousands of these structures are being randomly built, despite no one seeming to inhabit them. Occasionally you see a single woman in a hut, sweeping the floor, with a few chickens around.

It's a paradox: after the Spanish invaders, we left the Spanish invading reed, which in turn allowed the native Peruvians to become *invasores* in their own land. Once upon a time, Spanish, French, and English colonialists would arrive, stick a pole with a national cloth on it into the ground, and claim the land. Now all you need is a few stems and you can claim whatever you can, with no regard for the land around you. No cloth required.

❧❦

Of the many wonderful characters I met in Peru, there is one that always comes to mind—Félix Quinteros, who has known one of my colleagues at Kew, Oliver Whaley, for twenty years.

Félix was born in 1952, in Comatrana, a village of farmers and fishermen a few miles from Ica, and spent most of his life planting trees there. He remembers life when he was eighteen, when there were still huarangos growing wild nearby and at the foot of the dunes, and how the villagers collected their branches and pods for animal fodder. He told me how the houses of Comatrana were constructed with *horcones* (thick beams of huarango), cane, and mud, and how the roofs were a mixture of mud and donkey manure. He remembers the heavy winter fogs that allowed *Tillandsia purpurea* (known as "air plants," a relative of the pineapple) to grow in the huarango forests. They called those times *la blandura* (the softness) because of the dampness that soaked the earth. The groundwater was so high, it kept the crops and vegetation of the pampas alive during drought.

There was more water around then. There were still the lagoons La Victoria, Saraja, and Pozo Hediondo, where Félix went to bathe with his family and learned to swim despite the mosquitoes. Félix's great-grandparents said that where a huarango grew inside an animal pen, there were no pests or diseases: "That freshness, that aura, that energy, that scent of the huarango prevented the diseases of the ageing animals," he told me. These animals, in turn, protected the huarango from diseases. Chickens and turkeys, for example, ate the "leaf-picking worm," or soya bean webworm.

As a child Félix climbed the huarangos and would swing on a rope between branches. He played *gallito* with the other children—a game that consisted of finding the longest and hardest huarango pod and throwing it like a blade into the sand. If it stayed upright, you could give your opponent a *cocacho* (a crack on the head with your knuckles).

The people of Félix's generation no longer wanted to work the land or raise animals. They migrated to the city of Ica or Lima in search of a better income. They wanted to be builders, mechanics, and typists. As they left, Félix witnessed the destruction of the trees he loved the most: the mass logging of huarangos.

Many times he urged relatives and neighbors to replant them. But nobody listened. It was then that he began to take photographs of their demise, as a way to channel his pain and to protest. After seventeen years he had his first exhibition of photos in the Plaza de Armas at Ica, using borrowed painters' easels and paperboard holders to display his pictures.

People made fun of him. They called him *huevón* (a "silly dude" in Peruvian slang) and *loco* (crazy) for photographing something as insignificant and humble as the huarango.

"You could have at least photographed something nice, like avocados, orange trees, or grapes," they said.

The experience left him shocked and upset, though he quietly continued to document the huarango. He did not know why he did it; he just knew that he was drawn to trees and plants. Félix later studied agronomy (the science of soil management and crop production) at the Facultad de Agronomía de la Universidad Nacional San Luis Gonzaga (UNICA), and then began working for them. He proposed to the board that they should plant huarangos, but his idea was rejected. In an act of individual protest he started a nursery at home and gave away the seedlings. Soon enough he became known as "Huaranguito," or "Little Huarango."

Félix now campaigns tirelessly for the huarango. He lost his home in the port city of Pisco during the earthquake of 2007 (which killed 519 people), but that has not stopped him. He still nurtures the trees and plants them. Though Kew's and ANIA's (Asociación para la Niñez y su Ambiente) nurseries have so far planted at least 100,000 trees, only one in ten are still alive in the desert, as it is difficult to establish young plants with shallow roots when wild animals graze on them. In the face of this adversity, Félix forges on. He hopes his teachings will inspire thousands of children to plant a huarango tree.

In the villages we traveled to, Félix ran sessions on how to harvest, extract, dry, store, and sow huarango seeds. It is trickier than you might think. The pods are hard, and the seeds small and well encapsulated, but Félix has developed a method to get to them using a bladed can opener and a nail—simple tools that are widely available. He opens the pods faster than anyone else.

I provided sessions on how to take cuttings, prick out seedlings and graft plants, and Félix would then give a speech: "These people, the Kew team," he said, "they are different. They don't come here to take from you—they work hard and want to help. They're not here for money, like others. They're here to give you resources, support, and education so you can save the forest. It's up to you now. No more lying about complaining, or playing football and getting drunk. You need to change, work for your forest, and plant a future."

Most men in the villages knew that he was right, and the truth was like a wrecking ball. In every speech he makes, sooner or later Félix says: *"El mundo no es difícil, lo hacemos difícil"* ("The world is not difficult, but we make it so"). This motto has rung true for me ever since.

❧❧ ❧❧

Neoraimondia arequipensis, mentioned briefly at the start of this chapter, is the world's largest multibranched shrubby cactus—its angular, starkly beautiful, and rather impressive stems are an iconic sight in the areas in which it grows. Individual stems are about sixteen inches in diameter, and are sixteen to thirty feet tall. The spines, in clusters of up to seven, can grow to ten inches long. The flowers—which are reliably produced—are typically light pink to light yellow, but fruit generally develops only when there is water. The fruits are red on the outside and white or purplish-pink on the inside and packed with tiny seeds—like a small dragon fruit (*Pitaya*). This cactus is endemic to Peru and is found in the central arid plains from just above sea level to an altitude of 9,000 feet.

These plants are magnificent enough when admired from afar, but if you get close and are lucky, they might talk to you. No, this is not like the hallucinogenic peyote or San Pedro cactus—I'm not talking about opening doors into the afterlife or a drug-induced trip. I'm referring to what's known as "cacti graffiti," or, if you prefer, "cacti tattoos."

Using a pointed implement, you can carve marks or words into the cactus skin that will never heal. Normally I would say this is unacceptable plant abuse—even sacrilege. But in this case I make an exception.

Oliver Whaley, from Kew, has spent more than twenty years traveling and working in Peru. He showed me some incredible examples of this kind of cacti graffiti around Ica. The remarkable thing is that much of the writing is dated and occasionally offers valuable information, especially about the weather and changes in vegetation. The earliest one he has found dates from 1902 and corresponds with the arrival of a preacher, who would have taught people to write. Most are in flowing calligraphy and record moments of significance to the local area—especially the annual arrival of river water: "1934 . . . waiting for the water."

Some are romantic: "*AMOR, Cuando por las mañanas te despierta el aire, no te asustes, por que es un suspiro mío para ti, ROSA*" ("LOVE, when in the mornings the wind awakens you, don't be afraid, because it's a sigh of mine for you, ROSA"). Others record moments in people's lives, some simple—"I was hunting pigeons . . . food, 12 April"—and some highly descriptive—"10 January 1975, last day of work, going to Lima to enroll in the army while passing here this afternoon, writing in the presence of Don Ezequiel, already 60 years old, having had his birthday on 10 April, and Hipolito . . . born 13 August 1960, today 17 years old, eligible for military service." One I particularly like, etched carefully near the bottom of a large cactus, is: "I have eaten all the figs in my orchard" . . . straightforward and to the point!

It's hardly surprising that *agua* (water) seems to be the most common word in the cactus scripts—one from March 1921 records *"Vino un buen aumento de agua colorada"* ("There came a good increase of red water"). This would have been because the water from the Andes is often loaded with colored sediment. The dates of the arrival of river water provide us with insights into hydrological systems and the corresponding plant phenology (cycles and seasonal phenomena). For example, the inscriptions show that between 1917 and 1957 the water tended to arrive during January or early February, whereas today it arrives later, about April.

Cacti are like nature's time capsules in the desert. If someone who walked there before thought something important enough to leave a message about, we should do our best to learn from it.

❖

The National Reserve of San Fernando, about half an hour's drive from Nazca, is devoted to the rich wildlife areas of the sea and the fog-dependent vegetation by the coastal hills. It is one of the best national parks in Peru for biodiversity, boasting 90 species of desert plants, 90 species of fish and crustaceans, 252 species of birds, and a host of other animals and reptiles. The Humboldt Current flowing along the coast is relatively cold, like the sea in early summer at Brighton beach, and cools the air above it. This spreads over the hot land at night, and when the hot and cold air collide, fog or mist forms, which waters the plants.

I and the rest of the team, including Oliver Whaley and William Milliken from Kew, left the main road at night and arrived in the vast desert that runs between Nazca and the coast as the sun rose. There was hardly anything alive. Our shadows were many feet long and projected themselves endlessly over the dunes as we drove.

There is nothing along this strip of the coast for about 200 to 250 miles, and in some areas it is not thought to have rained for thousands of years. Our plan was to cross the desert through the center, reach the coast, then drive from north to south before returning to Nazca. We started at 4 a.m. to avoid the midday heat of the desert, and only got to bed at around 1 a.m. the next day. With no stop for sleep, we drove for more than twenty hours straight.

The red desert, with its weathered stones and sand, was how I imagine Mars. It felt like we were driving over ground that had never been disturbed before, but after a while we found some tire tracks and decided to follow them to avoid further damage to the desert. All around us there was nothing.

We drove up a mound about sixty feet tall to survey the landscape; in the distance we could see ripples in the sand. As we came closer the ripples became lines of *Tillandsias,* known as *clavel del aire* (air carnation) or *clavelinos* (little carnations) in South America. They can survive in the harshest conditions imaginable, where even cacti do not grow. All the lines faced the same way. The wind blows through the *Tillandsias* and the angle of the leaves makes ripples in the sand; as the ripples increase in size, the *Tillandsias* grow to compensate. The bodies of some *Tillandsias* have been found to extend ten to thirteen feet down into the sand. Thanks to the results of carbon dating, they are believed to be growing from a seed that germinated 14,000 years ago, making them older than "Methuselah"—a Californian bristlecone pine that is nearly 4,900 years old.

When we looked closely we could see the condensation on the *Tillandsia* leaves; behind each plant, in the shadow, the ground was wet where the water had dripped from the leaves. It was here we discovered a few specimens of a member of the potato and tomato family—*Solanum edmondstonii,* found nowhere else in the world. It is not your usual member of the potato family; it has small leaves, about an inch long, that are similar in shape to an English oak but blue-gray in color. The flowers, meanwhile, look like poppies from afar, both in shape and texture, and are pale mauve fading to white. Even though it was almost impossibly dry, it was still able to bloom, all because of the fog.

⟫ ⟪

We were deep into the desert, miles from anywhere, when we came across a giant PVC pipeline, about six feet in diameter, which had been laid out across our path, waiting to be connected by a mining company. We had to make a decision: should we go left or right? If we went in the wrong direction, we might not have the fuel to get back around the correct way.

We gambled and headed left. After driving for a good forty-five minutes with no luck, we found some workers in a jeep. They told us they did not have the cranes to move the massive pipes, but if we drove in the opposite direction for a further hour and a half, we would reach the main mining camp, where they'd be able to let us through. Imagine how frustrating it was to be in a vast untouched desert and to meet a barrier that couldn't be crossed, even in a jeep. We finally continued toward the coast, but time was ticking.

In the dunes we found stalks and seed heads from plants in the lily family, as well as other annual plants such as *Nolana*, a member of the tomato family, which go from seed to flowering and then die, all in a six-month season. We'd missed the blooming season, but the target was to collect seeds and put propagation procedures in place locally for the restoration of habitats and conservation. We also collected herbarium specimens to add to the detailed picture of plants from the coastal hills of San Fernando that we were building up at Kew.

One type of plant that particularly stood out was the endemic *Ambrosia dentata*, a relative of the Peruvian ragweed and a member of the daisy family. Normally it would grow in the mountains, but here it was growing on the coast. Its seeds have hooks like Velcro, and they must have latched onto the coats of guanacos, which are wild relatives of the domesticated llama. The area has a small population of guanacos that come all the way down to the Pacific coast from the Andes in the winter to escape the cold and feed on the vegetation. In ancient times the population was larger, but now it is believed there are only six left. For one species at least, guanacos are still important for seed dispersal.

As I was studying the plants, I caught a waft of ammonia—the kind of odor that makes you cough and your eyes water. In other parts of Peru we'd discovered chicken farms by the coast. It smelled like that—only worse. I thought I would wander to a vantage point on the cliff, where I could find out what was creating it. The landscape was such that you could never see anything of the beach until you were right at the edge, when suddenly everything would be revealed.

Down below was utter mayhem. There was a colony of thousands of South American sea lions, squawking, fighting, and giving birth. They were the cause of the foul smell. In the waves, dolphins were jumping around, playing games, and on the beach and in the sky there were Nazca boobies, Inca terns, Neotropic and red-legged cormorants, and Humboldt penguins. The contrast was astounding: behind me, the land was a lifeless desert; before me, the sea and shore were teeming with life.

I managed to get down onto the beach. The noise was almost deafening. Some of the sea lions were absolutely humongous, not to mention smelly. The thing that really amazed me were the condors, which, along with the red-headed turkey vultures, kept the beach free of disease and rotting carcasses. These condors, with their ten-foot wingspan, were breathtaking to watch, controlling their flight with a twitch of a wing or tail, swooping down in an instant. Some would have come down from the mountains, but there is a permanent population nesting on the coast because the food is plentiful—providing you like sea lion for every meal, that is.

❊❊❊

We left the San Fernando National Reserve and headed back for the last day's training. After this and a few botanical surveying trips, the expedition would be done. I was staying in Copara, a small town in a valley called Las Trancas in Nazca, and was teaching the people about propagation, reforestation, and food crops. The area is dry but they are lucky to have water from a nearby river and reservoirs. The town had been shaken by an earthquake a few months before, leaving nearly all of the buildings leaning at precarious angles, including the town hall, where we were staying. It had to fall sometime.

One of our tasks was to map the vegetation of a small side valley, which was shallow like a saucer. It was reached by a toll road—the barrier being a piece of string. One end was held by a guy sitting in a shed; the other end was tied to a post. The string lay across the road most of the time, and he would just pull it taut when he saw you coming. It was his job to count you in and out, day after day, the same time, the same man. The traffic was minimal.

It rains there only every ten to twelve years, during El Niño periods. The rains had been expected two years before we arrived, but they still hadn't come. Everything was parched and dusty, mainly beige rock and rolling dunes, but there was a welcome ribbon of green in the center of the valley, where there was also a dry riverbed. There must have been a supply of underground water. Away from the riverside, even the cacti were struggling.

When we were working higher up the valley, I noticed a system of terracing made of rounded cobblestones from the nearby river bed. If it were anywhere else in the world, I would have suspected they were rice paddies, but here? The people had evidently gone to a lot of trouble to make them, but there was nothing growing in them—they were filled up with sand.

The local guide told us they were probably used to keep sheep in, long ago.

I was bemused. Why would you want to keep sheep here? There were no plants, no water, the midday sun was ferocious, and the little vegetation that existed in the bottom of the valley—far away, down the hill—could barely keep itself alive. There was no trace of human activity at all. As I botanized, I kept thinking.

I then found some partially buried corn cobs, about one to one and a half inches long. They were obviously a wild species—I knew that in the pre-Columbian era they had been much smaller. But no one would have eaten cobs like these for the past 200 years. I dug around the area where the cobs were and unearthed a piece of colorful crockery with a depiction of an eye crying blood—just like a piece of pottery I had seen in a museum in Ica a few days earlier. The pot in the museum depicted a Nazca goddess crying rivers of blood, with croaking toads as earrings. It's not the kind of thing you forget easily.

I kept sifting through the sand and found more ceramic pieces. On several shards there were images of women, representing fertility and life. Everything was well preserved because it was so dry. This had to be an ancient archaeological site.

Then something caught my eye. It looked like a human tooth. The guide dismissed it, saying it probably came from a guanaco, but I thought it was more similar to my own teeth than to those of any other mammal I knew. I continued excavating and found a headscarf wrapped around some long black hair. I stopped.

"Do guanacos wear headscarves around here?" I asked.

Nearby I found a lower jaw, obviously from a human. I was stunned. This must have been a sacred site, or a cemetery. I immediately stopped my excavation.

I had seen the mummified remains of the Nazca people in museums. They were buried hunched up in a sitting position, with their knees by their jaws, wrapped in layers of blankets. Some of the skeletons had unusually shaped heads: when children were born to the ruling classes, a band was wrapped around their heads to constrict their skulls, which grew upward and formed a cone shape. It is called "artificial cranial deformation." They also practised trepanning, drilling one- to two-centimeter holes in heads to treat mental illness—or, as they thought, to release evil spirits.

When we told the mayor, rather excitedly, he was unmoved. They already knew there were cemeteries in the valley there, and had counted four of them themselves. But when I asked him to tell me the locations, it became clear that I had discovered a fifth, and that this was the only one on the terracing.

He led us to a storeroom in the lopsided council-house building and opened the door. Inside were several mummified corpses. The staff there were in the process of developing a small museum and exhibition. The bodies were buried with their artifacts, including small bottles made from dried gourds and decorated with pink, salmon, and orange feathers. I was told they came from the males of a species of bird known affectionately as "cock-of-the-rock," which lives in the subtropical Andes.

Later that day, in a restaurant, a man came and spoke to me in Spanish. He wanted to sell me a mummified baby's hand in a small glass case that was purportedly 2,000 years old. In Peru, grave-robbers will sell you anything—probably even their own grandmother.

✦⟫ ⟪✦

The morning after our day with the dead, we woke up and had breakfast as usual. We were all set for another botanical survey at a nearby valley, but our plans were quickly shattered: the person we were due to meet had been in a car accident and was a bit bruised, with an undriveable car. We were stranded.

As soon as the words left our messenger's mouth, I glanced over his shoulder and took a look at the Andes in the distance.

"How long does it take to get a cab to the Andes?"

"About two hours."

At last. Those mountains on the horizon had been calling me, and this was my chance. Alfonso Orellana, our guide, called the park rangers in Ayacucho to see if he, Kew trainee Doris McKeller, and I could stay overnight.

"Of course," was the answer.

I would soon be soaring like a condor or galloping like a guanaco, from the coast of Peru to the highlands of Ayacucho.

The guys who turned up in the cab had wrinkled, weather-beaten faces, as if they had spent their lives at high altitude, and talked between themselves in a dialect that sounded similar to the regional language, Quechua. After a while, they began talking to us as if they were running through the safety procedure on an aircraft.

"If we take you, there are some things you will have to do," they said, and started talking about altitude sickness.

"You know it can kill you. If you start to suffer, we will have to bring you down straightaway. We will only take you if, one, you wear a chullo [a traditional woolly hat, with ear flaps, to protect you from the cold and sunlight] and, two, you chew coca leaves."

The second request might seem strange, but we understood their reasoning. We were about to undertake a journey that would take us from sea level to about 15,000 feet in just two hours—our bodies would undergo quite a shock. The coca leaves would help to combat the physical effects of this.

You need something alkaline as a primer to make the coca work. In Peru, they use the ash left after burning the crop quinoa, a grain grown for its edible seeds. To prepare the leaf, they hold the leaf stem, peel it off to remove the central vein, which would scratch your gums, put the primer inside, like a sandwich, then wedge the plug between the cheek and gums. I was not expecting it to be so strong. I thought it would be like having a drink, but it was more like knocking back two liters of espresso coffee. It was a massive hit.

Forty minutes into the journey, we were following a road that snaked through the foothills of the Andes when we rounded a bend; there, stretched out below, was a valley full of plants and flowering cacti.

As usual, I shouted to stop. Before the van had even slowed, I yanked back the sliding door and leaped out, speeding down the valley, bouncing from rock to rock and plant to plant, jabbering with delight. When I finally returned, one of our guides poked his finger into my back and said, knowingly, "No more coca for you until we reach altitude."

At 13,000 feet, the effect was different. When there's a lack of oxygen your body almost grinds to a halt—everything feels labored. We were told coca gives you more energy, but it certainly didn't feel like it. Kneeling down to look at plants took an enormous effort, and getting back up again took more still, even with the coca.

I scanned the horizon with binoculars and spotted someone walking slowly in the distance. I couldn't see what he was doing at first, but as he got closer I saw a bulge in his jacket, which looked strange. Looking in the other direction I saw some sheep, and when he finally got to them, he opened his jacket to reveal a baby lamb sheltered inside, which he reunited with its mother. That was obviously his job for the day. It must have taken him six or seven hours to get there, and he presumably took the same trek back.

※ ※

The lowland landscape in the rain shadow of the Andes is extremely dry, like a rocky desert. In many places, you can go for miles without seeing a plant. Sometimes you might see some sparse cacti and the ever-surviving *Tillandsias,* but that will be about it.

Higher in the Andes there are the high plains—the altiplano meadows. At first they appear to be dominated by a feather grass called *Jarava ichu,* but if you look closely in between the plants there is a Lilliputian habitat of hundreds of tiny orchids, lupins, microflowers, mosses, and lichens. Before you know it, what seems like ten minutes looking at plants in three square feet has become one hour. There were so many I had never seen before, I just couldn't help myself.

I saw *Buddleia incana,* a form of the butterfly bush, which grows at altitudes of 8,800 to 14,500 feet. It looks like a cross between a hebe and a boxwood, with short leathery leaves and stubby, deep terra-cotta-colored flower spikes. I also found a form of what was probably *Bomarea dulcis,* with big hanging flowers that looked like a *Lapageria rosea,* a type of bellflower (the guard told me that children of all ages eat the red-coated fruits of the *Bomarea dulcis* like sweets—everyone in our group tried some).

It was growing over a *Polylepis*, a genus closely related to *Sanguisorba* (or burnet), a plant that is often found in cottage gardens. Both *Polylepis* and *Sanguisorba* are members of the rose family, and though the rose family is usually pollinated by insects, *Polylepis* is unusual because it is pollinated by the wind. *Polylepis* is the dominant species of the high-altitude forest: it is usually found at about 13,000 feet, though it sometimes creeps up beyond the 16,000-foot mark, far past the usual limit of the treeline in the tropics. This is the highest naturally occurring forest in the world, and it is a forest in miniature. Many of the *Polylepis* are gnarled and twisted—they could be 500 or 600 years old, but are only six to ten feet tall. The name *Polylepis* derives from the Greek words *poly* (many) plus *lepis* (flake or scale), referring to the multilayered bark that is common to all species of the genus, which protects them from extreme cold.

Since wood is very scarce in those places, the forest has been overlogged. *Polylepis* forests now exist primarily in small, isolated patches, and though some people are trying to restore them, their future looks precarious.

⤜⤛

Back in the car I was happily gazing out the window at passing plants. On mountains you tend to see more and more plants the higher you ascend, until suddenly, at about 9,800 feet, you hit an altitude threshold. From then on, while plants are still fairly common, they get gradually smaller in size, before you finally reach the domain of the alpine varieties.

As we ascended the Andes, the sunset lit the landscape rosy orange and backlit the *Browningia candelaris* cacti. The locals call this species *la candelaria* (which means "Candlemas" in Spanish), due to its shape. These cacti sit on the edge of sheer cliffs, the branching stems loaded with *Tillandsias.* The effect is like nature's attempt at contemporary art.

When the driver announced we had reached an altitude of 13,000 feet, I had a brainwave. *Puya raimondii,* the "Queen of the Andes," an endangered species and one of the most magnificent plants in the world, grows at an altitude of 9,800 to 16,000 feet in Peru. It is a type of bromeliad that can grow twenty feet tall and thirteen feet wide, and after about a hundred years produces a seven-meter flowering spike of 8,000 to 20,000 blooms, which are pollinated by the largest hummingbird in the world. After flowering, it produces millions of seeds, then, exhausted by this massive effort, it collapses and dies.

I asked the driver whether he'd heard of this plant. He asked his friend what the name was in the local language, and his friend made a call.

"The good news is that there is one near where we are staying; the bad news is that it is half an hour away."

I was in ecstasy.

At first we could see only ten to twenty plants, but as we walked toward them more *Puya raimondii* came into view in the valley beyond. There must have been a thousand of them. None were flowering, but some were fruiting. I'd heard that sometimes *Puyas* get struck by lightning, burst into flames, and burn ferociously; it certainly looked as though several had been burned in this way. It looked as though some Andean god had dropped his eyeliner, as they protruded, black and monstrous, from the ground. On one plant a stalk containing thousands of fruits had broken off at the base and millions of seeds had spilled out.

I have never been more tempted to take seeds without a

permit. But I pushed my hands firmly into my pockets, turned, and left them to rot on the rocky ground. I did not collect a single seed. We scattered some around in the hope that they might survive, and Alfonso brought some down to the nursery to see if the plant could be cultivated in the lowlands. We already had some growing at Kew, which helped ease my frustration.

As the road weaved onward and upward, the light changed. We arrived at our refuge, 13,000 feet above sea level, and the stars loomed out to greet us.

As the altitude increases above sea level on the dry side of the Andes, you see seedlings and juvenile plants, isolated specimens and groups, and guanacos grazing on creeping, hairy cacti. Trees don't exist, apart from in the misty, moist areas dominated by *Polylepis.* Finally, you reach that great contradiction, the *Puya,* like a giant pineapple, sitting in a valley at thousands of feet above sea level. How did it even get there, and how does it survive?

One of the answers is *Patagonia gigas*—the giant hummingbird, one of four hummingbird species recorded as having pollinated *Puya raimondii.* It is about the size of a European starling or a North American northern cardinal and is found along the length of the Andes in mountain scrubland and forest. Its wings move at fifteen beats per second, its resting heart rate is 300 beats per minute and it needs to consume 4,300 calories an hour to survive. No wonder it loves the nectar-rich *Puya raimondii.* I was told that if it is raining in the high Andes, this bird comes down to the lower levels because it knows the rivers will be flowing and the flowers blooming, meaning there will be plenty of nectar for food.

Even though *Puya raimondii* flowers once in a lifetime and blooming plants are not that common, you would still think that a plant that produces 8 million to 12 million seeds would be successful. But bad weather at the time of seed dispersal, poor pollination, a lack of suitable areas, and overgrazing mean that relatively few *Puya raimondii* make it to adulthood. If conditions are not right, the seeds lose their ability to germinate within a few months. The plant flowers only once, making it quite possible that a hundred-year-old specimen may not produce any successful seeds and so will live in vain. Some plants might also be set on fire by lightning just when the seeds are being shed.

In some years, especially in small populations, the plants will not flower at all; in other years, several plants will flower and set the tiny, heart-shaped seeds that are dispersed by the wind. Some will fall close by; others will be blown to another place where, if they are lucky, the seeds will germinate. It is random—most populations are dotted here and there and are several miles and valleys apart. One way of mapping *Puya raimondii* populations is by satellite. They look like pinheads on the landscape and can be counted individually.

Their growth requirements are specific too, so the chances of their finding somewhere new to germinate are small, and each group becomes isolated. This is proved by the fact that there is very little genetic variation within existing populations; DNA analysis of eight populations detected just fourteen genotypes in 160 plants. Four populations were similar in appearance as a result of inbreeding. Although the plant's genetic makeup is perfect for the harsh environment where it lives, it may not be flexible enough to cope with factors like the increased temperatures of climate change.

All of this means each group of plants shares a lot of the same challenges with plants growing on islands, such as *Roussea simplex* on Mauritius. As they form isolated populations and require a limited temperature range, there are limits on where they can grow. They stay isolated.

❖

We stayed in the Andes for three days. Physically it was highly challenging. I couldn't sleep; if I did, I stopped breathing and woke up gasping for breath. When we congregated in the morning everyone would be in a fragile state—we would sit and look at one another through squinting eyes for an hour and drink liters of coffee before anyone was in a state to talk. I was amazed there were no after-effects from the altitude (or the copious chewing of coca leaves).

There is something otherworldly about the altitude, the harshness of the landscape, the quality of the light, the drama of the Andes. Sometimes you see clouds coming, rolling along the surface of the ground. They crash into you and you are swathed in mist for a few seconds before they vanish. I think that's how *Puya raimondii* is watered—the leaves combing water out of the clouds. On that side of the Andes, at altitude, almost everything survives due to condensation from clouds and the rolling mists by the sea.

It's obvious that the Andes, with their receding glaciers, fog traps on the western side, and hyper-humid cloud forest on the eastern side, are the key to priming the ecosystems that exist there. They dictate what plants can grow and the crops that can be farmed at lower altitudes. If we don't preserve these unique habitats, everything will be destroyed on either side of the Andes and, with it, most of the ecosystems of South America, from the dry coastal Pacific forest to the lush expanses of the Amazon rainforest, the lungs of our planet, centered along the largest discharge of water in the world. It all starts here, high in the Andes.

※※※

With just a day before we were heading back to Lima, I still had one botanical wonder left to see: a horsetail, a plant that neither flowers nor sets seed.

They are the only surviving genus of a plant lineage that dominated the world during the Devonian period, from 419 million to 358 million years ago. They were a large part of the forest that ended up creating most of the world's coal deposits, which nowadays we burn freely. During ancient times some of their relatives were 100 feet tall; today, most species are much smaller, but there are still two South American species that fall into the category of giants: *Equisetum giganteum* and *Equisetum myriochaetum,* which can reach sixteen and twenty-six feet, respectively.

I asked Alfonso Orellana where I might find some specimens of *Equisetum giganteum.* Again, he got on the phone and came up with an answer. There was a population about two hours away; given how rare it was, it might be worth trying to propagate it.

The journey took us through miles of dry, dusty landscape, inhabited by just a few *Tillandsias.* It was becoming tedious, when out of nowhere we saw what looked like a rainforest, with giant trees and leafy growth. On one side of the car was parched desert, stretching out to the horizon; on the other side a lush rainforest. The green side was actually a huge organic farm. We had to pay it a visit.

Topara Organico Farm was created by Klaus Bedersk in 1968 as the first organic farm and nursery in Peru. In the valleys along the wide rivers that run from the Andes to the Pacific, he grows pecan nuts, fruit, and other crops. He also has boreholes, river water, and humidity traps high up in the hills to collect water from the atmosphere, to water his plants. It's like an oasis.

Klaus is a blond-haired, blue-eyed Peruvian, whose pregnant mother took the last ship out of Germany before Hitler came to power. His father started the farm, and Klaus eventually became the representative of Monsanto Agrochemicals in Peru. He turned to organic farming methods after a mudslide during an El Niño event destroyed every one of the farm's 350,000 plants. In rebuilding the farm, Klaus decided to switch to organic produce because he had no money to buy fertilizers and agrochemical products. In his first season, Klaus harvested just 80 percent of what he used to when using chemicals. Still, because it was certified as organic, the value of his crops was much higher. Profits went up, and he never looked back.

He also grows some ancient Inca crops, which the Spanish replaced with European crops such as wheat, barley, carrots, and broad beans in a kind of botanical colonialization. These ancient crops include *maíz morado*, a deep purple form of sweet corn, which is used to make a tasty drink called *chicha morada*. Then there's ají, a type of chilli pepper that was used as a flavoring by the Aztecs and Incas. He also grows lucuma, the "gold of the Incas," a subtropical fruit that tastes like maple syrup.

When I met Klaus, his main crop was pecan nuts. I soon saw that he was growing the kinds of plants that would flourish in the native climate and conditions, sometimes even creating his own crop varieties, and encouraging insects that devour pests by planting strips of wildflowers alongside. He also increased the organic matter contained in the soil by mulching the beds heavily; this not only fertilizes the soil but also prevents weeds and increases water retention. Rather than fighting the environment, he was working with it, and working sustainably, using solar energy and biofuels. He reckons his profit has increased by 200 percent.

Klaus has faced tough moments. In 1969, after owning his land for only a year, a left-leaning dictator came to power in Peru. The government seized a third of the farm and never paid Klaus for it. In 1980, the Shining Path, a terrorist group that killed thousands, visited his valley. They held civil trials that almost always ended in the execution of the wealthiest landowner. Klaus, the only white man in the valley, was put on trial, but his generosity toward his poorer laborers led him to be spared.

His farm is inspiring, living proof that things can, must, and should be different.

❧❦

Back on the hunt for *Equisetum giganteum,* we learned that it has a scattered distribution in Peru. In one location, we discovered the river had changed course during the floods and had swept the plants away in mudslides. We finally found the plant a little distance away, deep in boggy mud.

It was a spectacular sight, about 15 feet tall, with five or six full-sized stems—the rest were small runners that were starting to pop up, or larger stems which had lost their tops or were kinked. In some parts of the riverbed we sank down in the mud to our knees. I felt as though I was in the Carboniferous period, surrounded by silt and a few sedges and aquatic plants, waiting for a dinosaur to appear. You would not expect to find a horsetail, which is aquatic, living in the deserts of Peru; it survives only because the rain god occasionally cries. We collected some for the local nursery in Ica and some for Klaus's farm, where they have since flourished. In search of this botanical wonder, I had stumbled across a trailblazing organic farm. The world of plants is full of surprises.

14

Flora of Australia

Australia is not just the land of the platypus and koala—or, indeed, glorious waterlilies; plenty of its other plants have extraordinary stories too.

Drakea, the dragon orchid, has a flower that mimics the female of a particular wasp species in shape and color, even releasing a sex pheromone to further attract the males of this species, who think they are mating with a female but are pollinating an orchid instead. Then there is *Rhizanthella gardneri,* an orchid that is a parasite on the roots of the broom honey-myrtle (*Melaleuca uncinata*), an evergreen shrub or small tree that grows up to six and a half feet tall and is similar to the European broom bush (the only real difference is that instead of having pealike flowers, its flowers are like small fluffy ivory pom-poms). The orchid flowers underground and is pollinated by termites and gnats.

Then there are some wonderful tree species. One of Australia's most iconic trees, *Adansonia gregorii,* a species of baobab, has a trunk shaped like a ball. This helps it to store water and survive the drought period, which lasts for about six months every year. This species is scattered throughout the grasslands and is seemingly everywhere in Kimberley, making it impossible to ignore. On the east coast, *Brachychiton rupestris,* or the bottle tree, is shaped like a wine bottle, even down to the narrow neck, and looks spectacular dotted through the landscape. You feel as though you are driving across a giant's dinner table.

The diversity of eucalyptus species in Australia is beyond belief. There are some astonishing acacias too, particularly *Acacia dunnii*—native to the Kimberley region and part of the Northern Territory. Also known as the elephant-ear wattle, it has the biggest "leaves" (technically they are phyllodes, or flattened leaf stems) of any acacia in the world, up to eighteen by twelve inches. These "leaves" are covered with a white mealy substance to reduce water loss and negate the effects of the sun, and it has funky, bright yellow, ball-like flowers. It is found in open savannah forest, growing with spinifex (or porcupine) grass. The latter makes cushions of foliage and binds the desert together thanks to the roots, which go down ten feet into the ground. The leaves look soft and fluffy from a distance, and you might think you want to run and jump into them. Doing so, however, would be a big mistake—they are very tough. In fact, the sharply pointed leaves contain so much silica to keep them rigid that the blades snap into pieces rather than bend. Spinifex is extremely drought-resistant and can thrive in poor soils. It's clearly not a softie, despite its looks.

❊❊❊

I have to admit I have a special fondness for Australian water-lily species. A few years ago they were considered very difficult to cultivate, but this did not stop me from trying. With a combination of interest, trial and error, patience, and the facilities we have here at Kew, I worked out their basic needs and succeeded in cultivating them. Once I had mastered growing Australian waterlilies, people started sending seeds to me. At Kew I currently grow a dozen species and even have several forms of some of them from different localities, varying in color, size, and sometimes bloom shape. There are now about sixty kinds in the living collections, all growing under the same controlled conditions, where I can note the differences in shape, form, and color and observe their behavior, as well as use them for display.

I built up contacts with waterlily experts such as Andre Leu, who knows all the best locations for waterlilies in Australia; U.S. botanist C. Barre Hellquist, who regularly travels with Andre; and Emma Dalziell, who was doing a Ph.D. on the possibilities of safeguarding all the Australian species by collecting and storing seeds in seedbanks, overseen by Professor Kingsley Dixon, foundation director of science at Kings Park and Botanic Garden in Perth at the time.

The stars aligned. Kingsley received funding for a ten-day waterlily expedition in Australia. He thought it would help if I joined the trip, because I know about their classification and would be able to give them the correct Latin names and put them in the right plant groups. I would also be able to advise the gardeners on how to grow them, both in and out of the natural ranges where they are found. It was a trip I'd been dreaming of for ten years, and its scope was vast—it would be like collecting waterlilies in Portugal one day, then Turkey the next, followed by a stop in Italy on the way back.

There were seven of us. The first leg involved traveling through the Kimberley region of western Australia, one of the most isolated places on the planet, and also one of the last unexplored regions for Australian waterlilies. As Kingsley said, it was "one of the most biodiverse hot spots for waterlilies in Australia, and possibly one of the best spots in the world." He and his associates have spent thirty years categorizing about 4,000 plant species, and are still counting. When you realize that Kimberley is about three times the size of England, you can see the challenge.

It meant we would be taking the Gibb River Road, one of the toughest roads in the country, where flat tires and engine problems are to be expected, the water levels in the rivers rise and fall irregularly, and, of course, crocodiles guard the riverbanks. The Gibb River Road website advises visitors to "Be Crocodile Wise." For a newcomer, some of the advice may seem alarming: "If in doubt do not swim, canoe, or use small boats where crocodiles may live . . . Do not assume the absence of signs means that it is safe to swim . . . Be aware that a crocodile can remain hidden underwater for long periods and so you may not see it at all . . . Avoid the water's edge wherever possible . . . Don't paddle, prepare food, or wash . . . Don't hang arms or legs out of or lean over the edge of a boat . . ."—not exactly what you want to hear if you are trying to collect waterlilies.

I was then to fly to Cairns on the other side of Australia, spend a few days with Andre Leu, and travel up into tropical Queensland with Emma Dalziell, in search of more waterlilies, putting in more than 3,700 miles by jeep in twenty-five days. The purpose of both trips was to collect as many waterlily species as possible for herbarium collections. We wanted to record what species were growing and collect plants and seeds for storage in seed banks in Australia and for public displays, research, and germination trials at Kew.

Some of the plants we were searching for were very rare. *Nymphaea hastifolia* is a tiny species with pads shaped like arrows and starry white blooms that sometimes look a bit more like large white daisies than waterlily flowers; it has been found in just five or six areas in the Northern Territory and Kimberley, mostly in temporary peat bogs. *Nymphaea alexii* is a lovely species whose flowers are made up of perfectly arranged, pointy petals, with red centers and cream-colored stamens, and is known in just two lakes on the outskirts of a town called Normanton in Queensland. The best time to collect waterlilies is after the rains as the water recedes, when they are in bloom and you are likely to find seeds. As there is no fixed rainy season, you have to respond to the weather. Some lakes and creeks gradually dry up after the rains; others are permanent, making collection difficult to plan far in advance. This would be a testing trip.

As I flew in, I gaped at the wilderness of Australia, where the land has been left largely untouched by humans for millennia—so different from Europe, with its fields and roads.

The following morning I woke up in Broome, in Kimberley, after a few hours' sleep; our next form of transport—a jeep—was waiting. We hit the road and were soon in the great outdoors.

❈ ❈

Once you are in Kimberley, it isn't long before you notice the termite nests that are everywhere, scattered through the landscape. They can be up to thirteen feet high and six and a half to ten feet wide (they are narrow at the base and then grow fatter, just like the baobab). They are rock-hard, but when it rains the water makes them like modeling clay. So the termites build extensions to their nests when it is wet and then the sun comes out and bakes them hard again. Sometimes small nests pop up in the main roads—imagine hitting those at 60 mph. These and the baobabs make the scenery quite surreal.

Each water hole, creek, and river has a name (such as Dog Chain Creek, Dead Chinaman Creek). They form the reference points in the landscape, just like we might say "past the pub and left at the village church." But clearly at some point whoever named them ran out of ideas, and sometimes they are just called Tenth Creek and Eleventh Creek.

I was expecting to have to rough it in the outback but soon discovered that Australian camping wasn't quite as demanding as I'd thought it would be. The two jeeps were fitted with everything we needed: a water tank so we could wash our hands and clothes and make a cup of tea, and a fridge and a freezer filled with salmon, steak, and gin and tonic. The tents could be erected in minutes. It was more like glamping.

But I still had to learn where to pitch up. One night I heard a flapping and a strange high-pitched screeching; then small things started falling on the tent. Unable to sleep for the din, I crept outside with a torch, shining it into the branches of the trees. There was a roost of fruit bats with wingspans about a foot and a half wide, fighting, squabbling and—explaining the noise of things falling on my tent—throwing fruit and excreting. I had chosen to camp under their favorite tree.

Our route took us out and back to Broome, then northeast to the Dampier Peninsula. William Dampier, whom the area was named after, was a privateer from Somerset who turned to collecting plants; he became the first Englishman to explore parts of Australia and was the continent's first natural historian.

Near to the Dampier Peninsula is Lake Louisa—a permanent lake, right out in the middle of nowhere, that covers 600 acres when it's full. Although the water level often gets low, it never dries out and is therefore home for the westernmost population of *Nymphaea violacea*—the most widely distributed species of *Nymphaea* in Australia, generally found in shallow waters with mandala-like blooms that can be white, blue, or pink but are always heavenly scented. One of the aims of the expedition was to check the genetics of this complex species. Kingsley explained that you could reach the waterlilies only when the lake was receding, but that you also had to get there before it dried out too much. The only track out into the lake was over black soil, which is deceptive. It can look navigable, but sometimes you drive out onto it and the whole car sinks— you just don't see it coming.

Kingsley told me how, on an earlier trip, the car had become trapped; they ran out of water so they drank a bottle of champagne they were carrying with them, and a few hours later they were slurping water from muddy pools. It took a day and a half to tow the car from the mud. This time, with no margin for error, we decided we would hire a helicopter—the "big brother" of the type that is usually used for rounding up cattle.

⋇ ⋇

"Driver, take me to Broome International Airport," I joked as we jumped into the jeep and headed off to the helicopter we had hired to take us to Lake Louisa.

Because of fuel constraints we opted to take only three people to limit our weight. With me were Emma and award-winning photographer Christian Ziegler. Once in the helicopter, as we whizzed over untouched vegetation, following a mud track, I felt like Superman. In the middle of the dusty scrub, a teardrop-shaped lake covering about seventy acres emerged. We hovered over the lake for a while before the pilot slotted the helicopter among some eucalyptus trees and gently descended.

The last time Kingsley had visited this lake, the waterlilies had been growing to a depth of more than ten feet. I was shown a photograph of someone lying next to a waterlily stem that was twice as long as them—it was incredible. When we arrived this time, the depth was less than half that.

I walked for about forty minutes—about a half-mile into the lake—toward a group of pelicans floating in the center, with the water just above my knees, right into the *Nymphaea violacea* and *Vallisneria,* a kind of freshwater seaweed that has an ingenious pollination mechanism.

The female flowers have a flower stem like a corkscrew—it grows upward and when it reaches the surface the coiling screw pulls the three petals down into the water, just enough to create a slope but not to break the surface tension. The male flowers are released at the base of the plant and rise to the surface to float around. (I think this is the only plant that has a free-floating flower.) When they float by a female, the funnel formed by the water tension makes them float toward the flower, so they drop in and pollinate it.

I also found *Nymphoides beaglensis,* a species discovered in 1987 and known in just four locations, which looks like a water-lily but is in a different family. This species, like all *Nymphoides* I have seen, has two types of flower, which are both male and female—one has shorter stamens and a longer style, and the other has a shorter style and longer stamens. I wanted to find both forms, but all I could find was the one with the shorter style and longer stamens—it could be that even with 100 million plants in two and a half acres, they all originate from one plant. I did manage to find and collect a small seed pod with ten or fifteen seeds, but only one eventually germinated.

❖❖❖

Plant conservation is at the very heart of why Kew exists, and sometimes you need a little serendipity to help it on the way.

It was April 25—Anzac Day—and the fifth day of our expedition. We were driving along in a remote area of central Kimberley, about 150 miles west of Kununurra, hot and bothered after yet another puncture. All of a sudden I caught a glimpse of some waterlilies in a gap between thick shrubs and grass.

"Stop! Can we please stop?" I yelled.

After much chuntering from both the jeep and passengers, we ground to a halt. I leaped out and walked into the bush—there, in a long, narrow billabong, were hundreds of waterlilies. I waded in and started shouting for everyone to come over. The second I had seen the white and light-blue petals from the jeep, I knew it was a plant I had been growing at Kew for seven years.

Barre Hellquist had first collected it near Kakadu in the Northern Territory. I had instantly thought it was a new species, but he thought it was a hybrid of *Nymphaea violacea* and another, yet to be identified, species. Every time it bloomed, I looked at its round, perfect shape and said to myself, "I know that you are a new species. I know that you are a new species." Now that I had found it growing in a billabong 1,200 miles away from where it was originally collected, I was proved right, because the likelihood of finding a hybrid growing several hundred miles away from either parent would be billions to one. (Interestingly, a billabong is an oxbow lake or pond, left after a river changes course—the same habitat where *Victoria amazonica* is found.)

Kingsley said to me, "You haven't even taken the plant out of the water and you're saying this is confirmation of a new species."

"I can tell it's a new species without taking it out," I said. "I've looked at the plant at Kew every day for seven years: the seeds are the same color, shape, and size as this one here; the sepals are covered in hairs that make it look velvety, just like this. It's not scented—unlike *violacea*, which always has a distinctive fragrance, a combination of freesias and fresh apricots. And it's not self-pollinating like all the *violacea* I grow at Kew either."

DNA testing would prove it for me. And we had obviously found a hot spot, because over the next two days we found hundreds of other specimens in temporary billabongs in the sandstone alongside the Gibb River Road. Our new species was highly variable, in colors ranging from all white to deep shades of blue and purple. Some were much larger than others, and there was a stripy variation with green stalks and sepals, while another was brown with black dashes and black lines on the flower stalk and sepals. This variation is also found often in other Australian *Nymphaea* species.

❧❧ ❦❦

The Gibb River Road has hardly any bridges, which makes crossing the river very . . . interesting. Jeeps have a vertical exhaust like a chimney, so the depth of the river is not generally a problem, but the speed at which the water flows can be. Sometimes wooden poles are positioned across the river as depth gauges; at other times you just have to guess. The best way to cross a fast-flowing river is simply to drive fast with the windows slightly open, so you don't get trapped inside by the water pressure if the jeep gets submerged.

We attempted to cross somewhere near Charnley River Station, and the first jeep managed it with ease. But as our jeep, the second vehicle, hit the water, the engine slowed. The water rose rapidly up the doors, the nose dipped, the back tires lost traction, and the back end started slewing around. The jeep jerked forward as the tires bit again, but once more the vehicle began weaving and waves of water threatened to swamp us. After some nervy moments we finally reached the other side, to whoops of relief all round.

We were on our way to Lake Gladstone, the largest lake in Kimberley, covering 430 acres, which you might think would be easy to find.

Emma Dalziell had discovered it was home to a rare white waterlily. It looked like *Nymphaea carpentariae*, with its cup-shaped blooms held high above the water, and ovoid seeds, but this species was usually found hundreds of miles away. We'd been told to follow a track for twenty-five miles, then at a given point to leave the path and drive over fields. Kingsley, using a satellite phone while standing on top of a termite mound, managed to speak to local park rangers, who told us the coordinates—we went to the left, as they said, but by the time we had driven through the forest and hit another river, we were lost again.

After four hours we found it, concealed by the surrounding landscape, and I was soon deep in the mud, feeling a little uneasy. The water was murky, full of sunken branches, and every now and then flocks of birds would suddenly take flight, which would make me think of lurking crocodiles. I was alone in the middle of the water. It was worth fighting the fear: I found *Nymphaea violacea* in the shallows, together with a plant that looked like *Nymphaea carpentariae*. Both were cross-pollinating, creating a ring of sterile hybrids between the two that complicated the taxonomy even further. On inspection, the plant did indeed look like *N. carpentariae*. Everything matched it but its location on the map.

⊱⊰

As we headed to Wyndham, the oldest and northernmost town in Kimberley, we found what looked like some more *N. carpentariae*—though larger specimens than the ones we had seen in Lake Gladstone. This population had previously been recorded as being *Nymphaea macrosperma*, but although the overall plant size matched this species, I noticed that the flowers themselves were bigger, like *N. carpentariae*. There were thousands of these plants.

But that was just the appetizer for the next day. On reaching the top of a hill we saw a huge expanse of desert, dotted by hundreds of charred trees. Below were dead grasses and other vegetation, and right in the middle was Marlgu Billabong, a vast expanse of water that attracted thousands of wildfowl. Waterlily heaven.

There were thousands, possibly millions, of them there, in pink, white, and blue, stretching into the distance, as far as the eye could see. But I was puzzled. On inspection, some plants matched the look and the traits of *N. macrosperma*, while others were more like *N. carpentariae*, and there were even some that looked identical to *N. georginae* and all the forms in between. They were all growing together, cross-pollinating one another, and were very beautiful indeed. As I walked around another part of the lake where the water was receding, I saw that each waterlily was in a different shade of pink and blue, with a varying numbers of petals—not one was the same. Each flower was a different size and shape, and the leaves were different too, in shades of bronze and green. I started to come to the conclusion that *N. carpentariae*, *N. macrosperma*, and *N. georginae* were in fact a single, yet very variable, species.

One day you confirm a new species; the next, you discover that three species are probably one. My hunch was later backed by DNA analysis. If these three are designated as just one species, they would take the name *Nymphaea macrosperma*, because this was the first of the three to be described.

This discovery has implications for conservation too. There might now be only one, instead of three, species to worry about, but the broad diversity of each population means that every time a population disappears part of the gene pool may be lost forever, as each lake seems to have its own particular form.

The sunset at Marlgu Billabong was magical. The sun was reflecting on the vast expanse of water below and the clouds above, in shades of blue, orange, red, and violet. The lake glittered toward the horizon, and when a duck was startled and flew up into the air, thousands followed, wheeling in a huge arc like an evening flypast. At one point thousands of cockatoos flew over the waterlilies toward their roost in the surrounding trees. Like the macaws flying over the *Victoria amazonica* at dusk in Peru, I can close my eyes and see it now. It was the kind of place I would like to visit every week for the rest of my life.

It was not as peaceful as I thought it would be, though.

Emma started to become agitated at my wandering among the waterlilies. "Get out of the water!" she shouted. "I saw a crocodile here last year!"

❧ ❧

We hunted for several other species of the Nymphaeaceae family too. When I asked some Aboriginal children about the uses of waterlilies, one took me to his mother, who told me how they ground the seeds for flour. I showed her a picture of a plant on my phone—it was originally called *Ondinea purpurea*, but because of the results of DNA tests it has now been moved to the genus *Nymphaea*, making it *Nymphaea ondinea*. It is sometimes known as the "aberrant waterlily" because it is so variant—as opposed to "normal" waterlilies, it has only four petals; the leaves are wavy, grow under the water and sometimes have a glittery purple underside; and their few floating parts are shaped like a horseshoe—not like a lily pad at all.

The lady said that she had seen it, and that the tubers were the tastiest of them all—not only was it a kind of taxonomical and botanical delicacy, it was a gastronomic one too.

Emma and her colleague Matthew Barrett had collected some *Nymphaea ondinea* seeds by helicopter but warned me that they would be extremely difficult to grow. *Nymphaea ondinea* grows in flowing water, where the pH is very acidic—about 5.5. Lab analysis has shown there are no detectable nutrients in the water. Back at Kew I managed to germinate it about fifteen times but, as of the time of writing, none of the plants has reached maturity. Perhaps with some extra technology, such as a tank with CO_2 injected into the water, and a precise feeding schedule, we would be able to do this.

It seems to behave like members of the family Podostemaceae (riverweed), which are spread throughout the tropics but only in fast-flowing water, like rapids and waterfalls. Many species in the Podostemataceae family are so rare they are restricted to small areas, some to a single waterfall. They grip onto rocks, not compost, and grow in water that is so clear and pristine, no one can find even the tiniest trace of nutrient in laboratory samples. But they must accumulate a minute, almost homeopathic level of nutrients by filtering them from the millions of gallons of water that flow past them each day. That is nigh on impossible to replicate in cultivation. Many species like this have already become extinct because they are so sensitive—it takes a lot of work from someone who really cares to protect them. We need to raise our game and develop ways of cultivating them or we will lose many more.

I flew to Cairns on the east coast of Australia to meet Andre Leu, brimming with knowledge about Australian waterlilies. He, along with botanists such as C. Barre Hellquist and the late Surrey Jacobs, had not only pinpointed the locations of many Australian waterlilies but had also previously discovered several new species.

I did not sleep on the flight. The plane landed at 4 a.m., Andre picked me up at 5 a.m. and by 6 a.m. I was once again up to my chest in water, looking at waterlilies. He knows every pool within 400 miles. On the journey from the airport we made eight collections containing five different species, including one from the drainage ditch at Cairns Airport—*Nymphaea nouchali*. It must have been a recent arrival, because it has no close relatives in Australia. It is bigger than *Nymphaea thermarum* but still quite small, spreading up to sixteen inches, and it normally lives in brackish water by the coast in Southeast Asia.

Cairns is not hot and dry like Kimberley, it is hot and moist, rainforestlike, and the humidity hits you in the same way as when you enter the Palm House at Kew. Whenever I go to a rainforest, anywhere in the world, it always feels like the Palm House—hot as a sauna, with a deep organic scent.

Queensland is rich in flora and fauna. Along the eastern side is a series of mountain ranges called the Great Dividing Range. The eastern side of this range has a moist, warm, temperate climate, while the west, in the rain shadow, consists of hot, dry savannah and eucalyptus grassland. Farther north (on the eastern side) it becomes thick tropical rainforest with climbers and rattan palms—the wet tropics of Queensland are a World Heritage Site. The vegetation is so thick that when you stand on the path and put your arm into the forest, you can't see your hand at the end. Climb the mountains and the vegetation changes to montane forest, noted for its cool temperatures, high rainfall and humidity, and masses of epiphytes—plants that grow on the branches of trees but don't take nutrients from them. It is also home to a rhododendron with candy-pink flowers called *R. lochiae,* one of two in Australia (*R. viriosum,* which is similar, was originally grouped with *R. lochiae*) that are found growing on cliff faces and as an epiphyte in trees.

Andre's local knowledge was invaluable. We spent a day in the Daintree River and toured the mangrove swamps. These are some of the most biodiverse areas on the planet, and home to the spitting archerfish as well as Scarface, a fifteen-foot "saltie" dubbed the "King Crocodile of the Daintree River" who was something of a local celebrity.

One fascinating plant you'll find there is *Entada rheedii,* a giant liana known as the sea bean, with the biggest seed pods on the planet (well over five feet long). These pods fall into rivers and float out to sea, where they disintegrate and the seeds float on. That's why they are found throughout the tropics. There was a pod hanging nearby and I wanted to collect the seed, but the boat driver wasn't having it. It was the only one on the plant and he wanted to keep it to show the visitors, so he produced two seeds from his pocket and gave me one as a gift.

If you put them in water, they float; if you plant them, they do nothing. It's recommended that you file the seed coat to make it thinner, so water can reach the inside; I have tried doing this before, but they just rotted, so I decided to leave the one from this guy floating about in the waterlily pond at Kew. I hoped that eventually it would grow a root, but it didn't, so I tried the "scarifying" method again, by rubbing the outer seed coat with sandpaper to make it thinner, so just a little water could get through. Happily it germinated and is now growing. I'm still unsure if it needs another plant for pollination, but it would be phenomenal to have the pods growing at Kew as well as in Australia.

❖❖❖

High in the hills of Queensland there is a small water hole called Big Mitchell Pond. It is tricky to find because it is just off a long, straight road and well hidden by vegetation. Andre has been there several times and still struggles to locate it; I went back there with Emma and we had the same problem.

It is only fifteen by fifteen feet, yet it holds two exciting species. *Nymphaea immutabilis*, the first plant we found there, is one of my favorites from Australia—the sepals are green and blue and the outer petals blue and mauve, but the inner rows of petals are always white. It is incredibly beautiful. We also found *Nymphaea violacea*, some with blue flowers and some with pink. One of these plants had an interesting characteristic: the first-day flower was blue and the second- and third-day flowers were pink. It was really something. Though it was extremely tricky, I collected a great many seeds and some seedlings, and I am hoping one will eventually show that trait. So far they've all been pink—the rarest color in the pond.

These two species were joined by the bright orange, feathery blooms of *Nymphoides crenata*. Though the picture was heavenly, the biting mosquitoes were from hell.

A bit farther on is Mount Molloy, an old mining town thirty miles northwest of Cairns. There we found a cemetery with a pond behind it, surrounded by a eucalyptus forest. My first thought was that if Monet had lived in Australia, this would have been his Giverny.

At the edges of this pond there was a plant that looked like *Chasmanthium* grass from afar, but the little spikelets showed they were, in fact, rice. It is not a well-known fact, but there's a huge diversity of rice species in Australia. I came across two or three types in the wild—the grains were large and had the potential to be used in domesticated varieties; in fact, DNA tests suggest that modern Asian rice varieties originated from Cape York in Australia.[7]

Farther into the pond, *Nymphaea immutabilis* and *Nymphaea violacea* were coming into flower. The *N. immutabilis* in Big Mitchell Pond, which is five minutes' drive from this one on Mount Molloy, were purple with a white center, but in this pond every one of them was different. Some had dots in the sepals; others were pink or dark purple with a white center.

We found a couple of populations of *Nymphaea violacea* in other locations along the same road, and one of them was particularly notable. There were perfect black ruler-straight lines down the flower stems, as though they had been drawn with a draftsman's pen, each a different width. I was desperate to find some seeds, but they had all been eaten by geese. That's probably the reason why, after pollination, the forming fruits sink deep into the pond to hide away.

✻✺✺✻

Emma spent a bit of time at home sorting out the seeds from our collections, and after meeting up again, she and I decided to cross Queensland all the way to the Gulf of Carpentaria—the bay formed between the long peninsula horn of Queensland and the Northern Territory. We would then turn back south through central Queensland, eventually heading back to Cairns.

We drove through Daintree National Park, over the mountains and across Queensland, toward the north coast and the town of Normanton, where the type locations of *Nymphaea carpentaria* and *Nymphaea alexii* can be found. (As mentioned, "types" are pressed herbarium specimens of a plant from which the first written description of the species was made.)

We stopped at an artificial reservoir—no longer used, it had become a wildlife haven. Not only did we find *Nymphaea immutabilis*, with larger flowers than we had previously seen and deep pink specimens, we also came across the carnivorous aquatic plant *Utricularia aurea* (the golden bladderwort), with its mass of feathery foliage. It's a funny plant because it doesn't have roots; instead it just floats about, sneaking between other vegetation and feeding on aquatic animals such as mosquito larvae. Every now and then it puts out a bright yellow flower, a bit like a snapdragon, and tries to lure bees to pollinate it.

I found *Brasenia schreberi* too, which grows in Africa, America, and Asia; only later did I realize it was indigenous. The leaf stalks are attached to the middle of the leaf blades, which float like waterlily leaves do. All the submerged parts of the plant are covered in a transparent mucus—the Japanese cut the shoots, boil them, and eat them as a delicacy. Apparently it is highly desirable, but I am not convinced.

Though the reservoir was a boon for wildlife, it wasn't easy to traverse. I got into the water, but there was a sudden drop and I was up to my waist; then I tripped and had to grab a branch to stop myself from going under completely. I got out of the water, removed my waders so that I could swim if necessary, and tried again. This time, when I went a bit farther out, I didn't touch the bottom. The waterlilies were difficult to collect because the water was more than nine feet deep.

Since I had no footing, the only fruits I could grab were unripe. You have to open the fruit first and check that the seeds of *N. immutabilis* are black, dark brown, or olive green. If you open a seed pod and they are not ripe, you can leave it and the plant will heal, with just the loss of a few of the many hundreds of seeds the pod contains. In the end we collected just one pod, and so far I have managed to germinate one seed. Just enough to keep it going and reproduce it at Kew.

I was determined not to leave until I found the gorgeous *Nymphaea alexii*. It was first described in 2006 from collections in two locations: one fourteen miles south of Normanton, the other fifteen miles to the northeast of the town, in billabongs that are filled with water only during the wet season. The flower is white, with neatly arranged star-shaped petals, and the stamens and pollen are cream rather than yellow. The peg in the center of the stigma is pink, but on the second day the whole female part becomes bright red. It is endangered, and the kind of plant that could be saved by cultivation. So far it has proved tricky to germinate.

Emma and I were lucky enough to spot *Nymphaea alexii* in the location to the northeast of the town. We both had a gut feeling it wasn't safe to go into the water there because of the risk of crocodiles, but I didn't want to leave the lily behind. I started looking around. The lake was divided in two by the road, cutting the access between the two sides. Someone had also erected a fence to stop cattle from getting into the road, and near where we were standing, the trunk and branches of a fallen eucalyptus isolated part of the lake. Most of this side of the lake was very shallow, most of it up to your knees and in a few areas to your hips. There would be plenty of time to retreat if we saw any movement from a crocodile.

I said it would probably be safe; Emma disagreed. This is the kind of dilemma you often face out in the field. One moment you think it is safe or are prepared to take the risk; the next moment you don't.

I decided to run a further risk assessment. The water was like crystal, so crocodiles would lack the camouflage they have in murky or weedy water. The soil was gritty and firm, so you could move quickly and soon be out of the water. The water-lilies were not so thick that you couldn't see between the pads either. I threw some stones toward the plants and poked around them with a long stick. Eventually, we convinced ourselves that it didn't look like a crocodile danger zone. The geese and ducks that were there looked unafraid, and Emma climbed to her vantage point. I walked along the bank.

In I went. I got to the waterlily and grabbed several seed pods, one of which was just bursting—the perfect point for germination. A few waterlily tubers were detached from the soil, so I collected some of these for cultivation and herbarium specimens and then I was out. It took a matter of seconds.

After collecting the *Nymphaea alexii*, we thought it was worth heading 200 to 300 miles south toward Greenvale—a journey that would take us through the night. We ended up at a house where the track came to an abrupt halt, but we still hadn't seen a single waterlily. We decided to knock on the door and see if the inhabitants could help us.

An old couple opened the door, and we introduced ourselves.

"Hello," I said, "I am Carlos from Kew and this is Emma from the University of Western Australia. We have been told that there are waterlilies somewhere near here and we thought you might know where they are."

"Oh yeah, there are several places here. Hold on a minute," the man said, then jumped on a motorbike and disappeared.

Ten minutes later he returned.

"I have been to where I thought they were. There were definitely some there two or three years ago, but now I can't see any. Is this a small waterlily?"

I thought I knew which one he was talking about: *Nymphoides indica*. We were looking for a particular form of *Nymphaea carpentariae*. "Does it have furry petals?" I asked.

"Yeah, yeah, yeah," he said.

"Ah, sorry, that's not the one we're after. Any idea where we could find something that's way larger, with bigger pads and blooms?"

The man paused for a moment, then shouted, and a younger guy came around from behind the house. We discovered that he was Italian and had wanted to learn English, so had decided to go to Australia. Somehow he had ended up in Greenvale, population 150, in the middle of nowhere.

"They are asking for waterlilies," the man said. "When you went out last week, didn't you see some of the larger ones?"

"Oh yeah," said the Italian guy, "I know where they are."

He guided us to where the track finished, then we walked through the forest, up a small hill. At the top, we caught our breath and looked up. The view opened out on to an endless lagoon filled with thousands of the waterlily we were looking for, every single one white. They were as dainty and serene as a troupe of ballerinas. It was as though heaven had come down to earth and sat there waiting to be discovered.

✦⟫ ⟪✦

Our final mission was to find a newly discovered species called *Nymphaea jacobsii* in Lake Powlathanga, west of a town called Charters Towers in North Queensland.

The lake is large, covering about 800 acres, and although the water levels drop in the dry season, there is a permanent deep section. It is perfectly round, with a ring of *Nymphaea violacea* around the edges. Within it is a ring of hybrids between this and the new species in the center—*Nymphaea jacobsii,* described in 2011. All of the species, forms, and subspecies interbreed, which is why it takes years of research to make sense of it all.

Lake Powlathanga was very low. There were no waterlilies in flower, so we couldn't collect anything there. The locals said it hadn't rained for five or six years. Another nearby lake was completely dry—so dry that the bottom had been colonized by trees, shrubs, and grass. We found a group of Australian feral camels walking around.

Rainfall patterns are irregular there, and the plants had outlasted drought before, but each time it happens it is a threat to their survival. How long can they take it?

✦⟫ ⟪✦

I'm always trying to squeeze in one last burst of collecting. On our way back to Cairns Airport, we stopped at a slow-flowing river by the coast and collected two *Nymphaea gigantea* plants and a few hundred seeds. This species is a classic among the waterlilies from Oz. When Captain Cook arrived on the east coast of Queensland, his botanist, Sir Joseph Banks, collected two species of waterlily—this and *Nymphaea violacea.* Unfortunately the type specimen of the latter was lost, so we have no way of knowing the identity of the true *N. violacea* with accuracy. It doesn't help that this species is variable too. Now we are starting all over again, albeit with the advantage of being able to analyze DNA.

That is why type specimens are so important. The only solution left is to find the same population that Sir Joseph Banks collected that species from, if it still exists, or sample all the populations nearby and see if they are similar. This way, we would at least be able to establish which of the many forms of *N. violacea* is the true type. A few other species that were previously lumped in with *N. violacea* have already been identified, and I have a feeling that there are even more species to be split from it. This is the opposite to what happened with *N. carpentariae, N. macrosperma,* and *N. georginae,* which we realized were all one single species. For the *N. violacea,* since it is an extremely variable and widespread species, we need to carry out extensive sampling before we can reach a conclusion. And that is what we are doing. It is a work in progress.

Finally, after covering almost 5,000 miles by road in twenty days, with forty-eight collections of water-lilies from fourteen species, it was time to go home. I couldn't wait. I had a new waterlily species to play with and show to the world.

Epilogue

One in five plants is now believed to be threatened with extinction. Although I have been dubbed "El Mesías de las Plantas," in reality I and the staff in my "Noah's Ark" nursery at Kew—together with *all* the world's botanic gardens and nurseries—can't save this planet's plants on our own.

Anyone can be a plant messiah. You only need to have a spark of interest. That interest leads to knowledge, that knowledge leads to care, and that care leads to action. Just like Raymond Ah-Kee in Rodrigues.

Or like Francisco Rodríguez Luque, a teacher and keen amateur botanist in Spain who loves trying to conserve his local flora. On a plant-hunting trip he found a plant in the foxglove family (Plantaginaceae) on the shady, north-facing side of a collapsed cave. It was dry there, but the plant was clearly thriving where water seeps from the rock. He showed it to botanists and they pronounced it a new genus—a rare occurrence in Europe, which has been very well explored. They want to call it *Gadoria falukei*—*Gadoria* because it was discovered in the Sierra de Gádor in Andalucia, and *falukei* because Faluke is Francisco's nickname.

This case came to the attention of another Spanish amateur botanist, Julián Manuel Fuentes Carretero. He contacted me in 2016 via Facebook, saying there were now only fifteen of these plants left in the wild. He collected seeds with some of his friends, germinated them, and grew them to fruiting size. He sent me some seeds, and I put some in the Millennium Seed Bank at Wakehurst Place and grew others for public display at Kew.

The cost of this project was almost zero. I was able to grow the plant before it was even officially named, thanks to two enthusiasts from the natural world (and Facebook). As a result, *Gadoria falukei* will not become extinct in the near future. What we need now is a long-term backup plan.

This proves you don't have to travel to faraway places and fly around the world to find plants to conserve: they can be growing close to home. Why don't you go out and take a look around where you live?

You can go hunting for plants, either with friends or by joining gardening groups, horticultural societies, natural science organizations, or conservation charities. There is always a plant that needs a little help, or an animal that would benefit from habitat management. From alpines and rare mosses to seagrasses, there are roof gardens to be constructed and sand dunes to be restored, and frogs waiting for you to provide a pond. There is always, *always,* a way to enrich your environment and biodiversity.

You can join local organizations such as the Wildlife Trusts, or you can support national charities like Plantlife or the Woodland Trust or Sociedad Española de Biología de la Conservación de Plantas (SEBiCoP), become a friend of a botanic garden like Kew, or support organizations such as Botanic Gardens Conservation International. Your aid can be financial or practical—local groups often rely on volunteers to maintain habitats.

You may be more of a political person or feel passionately about what's happening locally. If so, get inspired by the people who formed Sheffield Tree Action Group to prevent the felling of street trees in 2016—including pop stars, professors, and pensioners—who protested through nonviolent direct action.

Teachers, enthuse your pupils about the wonders of the natural world. Parents, encourage your children to grow vegetables, build them a pond and let nature take over, get on a bus and take them to the woods or the sea or anywhere where nature reigns—it is cheaper than Disney World and way more colorful and fun. (Oh, and next time you vote, how environmentally conscious is the party you vote for?)

In your garden—or on your windowsill—you can grow endangered species. Take the chocolate cosmos (*Cosmos atrosanguineus*) from Mexico, which is now extinct in the wild and conserved through cultivation in gardens. By the early 1980s the only known plants were those at Kew, propagated from a single clone. Like the café marron, it did not set seed for decades, until a lady in New Zealand managed to gather a few seeds from plants she was growing and raise some seedlings. It took a hundred years, and a journey from Mexico to Kew to New Zealand, then back to Kew, before it finally hit the Millennium Seed Bank freezer at Wakehurst.

Or have a go at growing *Abeliophyllum distichum* (white forsythia)—a winter-flowering shrub that is endemic to South Korea and found only in several small populations there. It's still listed as critically endangered. Then there's *Tecophilaea cyanocrocus*, the Chilean blue crocus, which was thought to be extinct until a new population was discovered.

Out in the wild there are plants waiting to be conserved in the most unexpected places.

There is a cave on Rodrigues Island—if we shifted some soil

from there, would we discover a new species or rediscover an old one, like the *Lobelia vagans*, which appeared in a nursery pot on Rodrigues Island? After all, a *Silene stenophylla* (narrow-leaved campion) was found 125 feet below the permafrost of a glacier, and is 32,000 years old. It was surrounded by layers that included mammoth, bison, and woolly rhinoceros bones. That makes me wonder how many herbarium specimens of extinct species there are with seeds that may still be viable.

In 2016, the State of the World's Plants Symposium, held at Kew, revealed that species are being discovered—and lost—at an incredible rate.[8] About 2,000 new species of plants are discovered every year, and 2015 was no exception. Alongside the new species of waterlily in Australia, there was also *Canavalia reflexiflora*, a member of the pea family, which was identified and described by a Brazilian researcher looking at herbarium specimens at Kew. It has disappeared from its original location but has been found in another part of Brazil on a protected site, although that too is now being threatened by coffee cultivation.

Several of the new species found in 2015 are already believed to be extinct. The habitat of one, a forty- to fifty-feet-high tree from the dry forests of Ghana and the Ivory Coast, has been cleared for agriculture or destroyed by fires. The habitat of another, *Ledermanniella lunda*, a three- to four-millimeter-high herb from the family Podostemaceae, is now the site of a hydroelectric dam. Diamond mining has turned the river there cloudy and brown, which is a death sentence for plants of that family. It may well be extinct by the time this book is published.

I mentioned in the Introduction that plants provide everything we need—our food, clothes, medicines, and much more. Kew's 2016 report amply demonstrates what plants do for us. At least 31,128 species currently have a documented use for humans, animals, and the wider environment: 5,538 provide food, 17,810 offer medicines, 1,621 supply biofuels, 11,365 are used for materials, and 3,649 provide food for farm animals.

It's clear that plants do truly remarkable things. But recent research suggests we may have even more to learn about what they can do.

Mimosa pudica (*pudica* means "chaste, modest, pure, virtuous") is commonly known as the sensitive plant. If you gently stroke one of its leaves, the plant famously responds by shying away, folding her delicate leaflets in half, packing them tightly, then folding back the leaf petioles. In a series of graceful movements the foliage almost disappears from sight. This modesty makes good sense, stopping it from being damaged by rain or eaten by insects and grazing animals.

Monica Gagliano, associate professor of biology at the University of Western Australia, thought there may be more to this plant than meets the eye. So she ran an experiment in which individual pots of *Mimosa pudica* were repeatedly dropped onto soft foam from a height of six inches: just enough of a jolt to make them fold their leaves in haste. She repeatedly dropped the plants at five-second intervals to see at what point, if any, they would realize they weren't under threat and stop curling.

After a few drops, some plants stopped closing their leaves fully; then more and more of them stopped protecting themselves. After sixty drops, when she ended the experiment, they all remained completely open. As she noted, "They couldn't care less anymore."

Had the plants learned that they weren't under threat? Or could they simply have grown exhausted and had no energy left to curl their leaves? Gagliano, anticipating such questions, put some of the supposedly "tired" plants in a shaker and they instantly curled up again. A week later she tried dropping them again onto the foam, and the leaves stayed open. Even after twenty-eight days, when she repeated the experiment again, the plants remembered that they were not under threat when dropped in this way.

But how, without a brain, could these plants retain memories? As she wrote in May 2014, "What we have shown here leads to one clear, albeit quite different, conclusion: the process of remembering may not require the conventional neural networks and pathways of animals; brains and neurons . . . may not be a necessary requirement for learning."[9]

Another plant that fascinates kids of all ages is *Dionaea muscipula,* commonly known as the Venus flytrap, which catches insects and even frogs. It developed this strategy because it grows in boggy ground where nutrient levels are poor—it's the only way this plant can get a decent meal.

The iconic "eyelashes" on the traps increase the surface area of the trap without adding weight, enabling them to catch larger prey. (Why grab a snack when you can trap a banquet?) The movement is not just the first "snap": after capture the sides move again, pressing against the prey to stop it from escaping. The plant then releases enzymes to dissolve its prey, often while it is still alive.

This is certainly remarkable, but scientists have discovered something else: the Venus flytrap can count.

For the trap to work, the prey has to touch the trigger hairs on the inside—not just once, but twice within thirty seconds. No one is quite sure how this "nervous system" works, and how the plant effectively keeps time and holds the memory of the first touch. All we know is that the plant converts those movements into electrical impulses, which reach its digestive glands and motor tissue. If nothing moves inside after the first touch, then the trap will open again: false alarm. But if it is an insect, it will struggle to escape—triggering more signals—the trap will close tighter and the digestive enzymes will start flowing. The more the insect moves, the more enzymes will be produced.[10]

There is plenty of communication going on underground too. Believe it or not, plants can interact via vast webs of underground fungi at their roots. This "wood wide web" weaves its way through forests, gardens, and arboreta. Nobody knows quite how far some of these networks might go. Plants use this fungal network to help out their neighbors by sharing nutrients and information. Mature trees support seedlings and younger specimens in the same way that parents care for children: seedlings growing in shade, for example, where food is scarce, receive more carbon from donor trees than those growing in favorable conditions.[11]

But plants are not always happy to accommodate all comers. Some, like the tree of heaven (*Ailanthus altissima*), deploy chemical warfare, exuding chemicals from their roots into the underground web and the surrounding soil; others, like eucalyptus, add volatile oils to their fallen leaves, permeating the soil to create a toxic barrier and prevent other seeds from germinating. They are nature's weedkillers.

Plants do not have brains or nervous systems as we understand them, but they nonetheless manage to communicate and respond to stimuli. They receive information, translate it, then respond; they attract pollinators and use natural phenomena for reproduction and seed dispersal; they host bacteria to provide nutrients, and establish an Internetlike network through fungi in the soil. In a single leaf there are constellations of millions of cells, each delivering messages at a chemical level. Researchers are discovering that it is through these messages that plants actually "talk" to other kingdoms, such as insects. This is not magic or witchcraft but another frontier of knowledge—and one that we have barely begun to unravel.

Each gene is a word; each organism a book. Each plant species that dies out contains words that have been written only in that book. If a plant species becomes extinct, one book is lost, and with it the words and messages it carried. We are burning the library of Alexandria every time we destroy a hectare of pristine habitat.

The Messiah performed miracles, including giving sight to the blind. I want to cure us of plant blindness. After all, show us a picture of a monkey in a rainforest and we see the monkey but not the vegetation that provides the monkey's habitat, or the medicines for the shaman (and perhaps one day for us), or the food and shelter for indigenous tribes. The image is actually one of biodiversity.

If it was up to just one person to decide how to treat the earth, only a fool would allow this to happen, yet collectively humanity behaves like a headless chicken.

So what can we do? There are many things to choose from, but here are my top three:

1. Stop burning fossil fuels.
2. Keep population growth at a sustainable level.
3. Harness the power of plants.

After all, plants are the only things in the universe capable of capturing and storing energy, and creating myriad different materials and molecules, while absorbing and locking up CO_2 from the atmosphere. What we exhale, they inhale; what we inhale, they exhale. They are the key to our long-term survival.

A forty-year study of African, Asian, and South American tropical forests concluded that they absorb about 18 percent of *all* carbon dioxide created by fossil fuels.[12] To get an idea of the true value of tropical forests, this means that they remove nearly

5.5 billion tons of carbon dioxide from the atmosphere a year, which should be valued at about $17 billion a year.

The solution to climate change needs to be radical. That's radical in the Middle English sense of "having roots" (from the Latin word *radix*, "root"). A global forest-management strategy is crucial, but we should also introduce an international ban on the destruction of pristine primary forests. Cropping and farming in ways that do not release lots of carbon into the atmosphere—for example, by cultivating the soil without plowing it, and letting cows graze on grass among trees rather than feeding them corn—also have a key part to play.

Remember that we have already done something similarly radical with commercial whaling, which is now banned. It looked bleak too when we discovered the hole in the ozone layer, but this is now beginning to close, thanks to bans on the use of chlorofluorocarbons in aerosol sprays and fridges. Do you miss them? Was it worth changing our behavior? Then change again.

I always think that looking out of an airplane window from high up in the air is like viewing the earth as an extraterrestrial. I like to take a window seat, and I turn the whole journey into a natural science experience. Sometimes I feel awe at the vast expanse of our planet—seeing snow-capped mountains with rivers running to the sea, vast deserts, and the aurora borealis. On my way to Bolivia I flew over Iceland and Newfoundland and millions of icebergs, before landing in a Miami swamp. When I fly over the lakes of North America I wonder if there are any waterlilies growing in them, and I imagine moose grazing by a glacier. Sometimes it makes me feel elated; at other times—when I see how the Amazon is covered almost entirely with smoke from forest fires—depressed. We live in a beautiful world, but it's in trouble. If we don't act soon, it will be too late.

Close your eyes. What do you see in the future: apocalypse or human societies changing their ways? Many see the apocalypse. We all need to start visualizing the change instead. Then we will change our attitudes and take action.

There was once a great debate on whether the earth was flat or round. Although many in ancient times suspected it was round, it wasn't until the sixteenth century, when we circumnavigated the globe, that we had practical evidence of the shape of the earth. Finally, in 1969, a spectacular photo taken by the crew of *Apollo 8* showed everyone just how round the "blue marble" was and how dark it was out there.

For those who study nature, the reality of climate change is as clear as the earth's roundness was to sailors. But this time, the final photographic evidence will be too late. Discovering the shape of the earth could wait; climate stability can't.

We won't find another planet and move to it: the chances of this happening are nil. We have been given one earth and we are not managing it properly. We don't deserve another.

Instead, let's turn things around and garden our way out of this apocalypse, green up the world, and plant our future.

Amen.

Acknowledgments

❧ ❧

I would like to thank my parents for nurturing my interest in nature from an early age, especially my mother, Edilia, who opened my eyes to the fascinating world of plants.

I wish to express my gratitude to the many people who contributed to this book, and for the stories and scientific papers they shared to make it come alive, and to acknowledge all the staff and trustees, students, volunteers, and associates who make the Royal Botanic Gardens, Kew, such an amazing place.

I am grateful to the late Ian Leese, principal of the Kew School of Horticulture, for helping me jump a few hurdles in my career, and to Kathleen Smith for her invaluable support during my early days at Kew (and later for ferrying me to and from Perth Airport, even at antisocial hours).

I would also like to thank Dave Cooke, manager of the Temperate House at Kew, for raising the first cutting of *Ramosmania rodriguesi* and for sharing his story; Dr. Paula Rudall for her generous help with this book, and for sharing vast volumes of research and information over the years; Oliver Whaley for introducing me to Peru and showing me its many wonders, and for the information he provided for this book; Alexandre Monro for the Bolivian training marathon and all the invaluable help with the Bolivian chapter; and Richard Barley, Lara

Jewitt, and Ciara O'Sullivan for all their support with this project and for proofreading the book too.

I wish to thank the following for the photographs and illustrations used in this book: Nigel Pickering, for the illustration of *Ramosmania rodriguesi*; Lucy Smith, for the illustrations in the glossary and the photographs of *Victoria amazonica* and *Nymphaea thermarum*; Alexandre Monro and Oliver Whaley for the photographs of Bolivia and Peru; and Dennis Hansen— *Roussoea-Trochetia-Chassalia;* and Christian Ziegler.

A very special thanks goes to everyone involved in the conservation of Mauritian biodiversity at the National Parks and Conservation Service, the Mauritian Wildlife Foundation, the Forestry Services of Mauritius and Rodrigues, and the Mauritius herbarium, for being so incredibly helpful and dedicated to the cause of conservation, and for their wholehearted efforts while working out in the field and in the nurseries. I am particularly grateful to Claudia Baider and Alfred Bégué for their input on this book; I appreciate your rapid replies to all my queries.

Special thanks to Mr. Kerry Stokes AC and Mrs. Christine Simpson Stokes of Wavelength Nominees for their generous support, which made possible the major Kimberley expedition and the discovery of the new waterlily species. Sincere gratitude goes to Professor Kingsley Dixon for his Herculean efforts in the organization of this collecting trip, and to Dr. Emma Dalziell for being my fairy godmother during fieldwork and for her input on the Australian chapter. To Andre Leu, Stephen Bartlett, and Lionel Johnston for their efficiency and organization and for helping to make my Australian collecting trips such enjoyable and productive experiences. Thanks are due also to the late Surrey Jacobs and Andre Leu for their pioneering work on Australian waterlilies, and to Professor C. Barre Hellquist and John Wiersema for their generous help over the years and their long-term dedication to increasing the understanding of waterlilies, and for revealing their diversity and beauty to the world.

To the Innocent Foundation for sponsoring the Bolivian projects and to Sainsbury's plc for their support of Kew's projects in Peru. A special thanks to the charities ANIA (Peru) and Herencia (Bolivia) for their unwavering support of Kew's life-changing projects in these countries; to the RBG Kew team in Ica and Lambayeque, Peru; to everyone in Conservamos Ica (www.conservamosica.org); and especially to Félix Quinteros from Peru for being such an inspiration—and such a dude!

To the gardening charity Perennial (www.perennial.org.uk) for all the incredible practical and financial support that they provide to horticulturalists in need in the UK, and for all their assistance.

To my editor, Joel Rickett, and my copyeditor, Caroline Pretty, for their insightful editing, and to the whole team at Penguin for believing in this project and publishing this book. I would also like to thank my agent, Jon Elek, and everyone at United Agents for their great help.

I am infinitely grateful to Matthew Biggs, my ghostwriter and collaborator in this project, for his patience and assiduous work. Without him, this book would not have been written in such record time.

To my partner, Genevieve, for her love; to my family for their never-ending support of my endeavours; and to my friends in England for being my family in my adopted country.

And finally to Jane Goodall for being the ultimate inspirational friend.

Notes

❖❖❖

1. Taken from C. M. Paulus Rey, "Incidencia de las juntas de extinción de animales dañinos sobre las poblaciones de lobo ibérico" (2000), https://www.scribd.com/doc/48146786/.

2. As distinct from *Flore des Mascareignes: La Réunion, Maurice, Rodrigues* (IRD Editions, 1999).

3. R. E. Vaughan and P. O. Wiehe, "Studies on the vegetation of Mauritius. I. A preliminary survey of the plant communities," *Journal of Ecology* 25 (1937): 289–343.

4. J. J. Rousseau, *Discourse on the Origin and the Foundations of Inequality among Men* (1755; reprinted by Indianapolis, Indiana: Hackett Publishing Co, 1992).

5. G. T. Prance and J. R. Arias, "A study of the floral biology of *Victoria amazonica* (Poepp.) Sowerby (Nymphaeaceae)," *Acta Amazonica* 5 (1975): 109–39.

6. R. S. Seymour and P. G. D. Matthews, "The role of thermogenesis in the pollination biology of the Amazon waterlily *Victoria amazonica*," *Annals of Botany* 98 (2006): 1129–35.

7. M. McCarthy, "Is Australia the home of rice? Study finds domesticated rice varieties have ancestry links to Cape York" (2015), http://www.abc.net.au/news/2015-09-11/wild-rice-australia-linked -to-main-varities-developed-in-asia/6764924.

8. https://stateoftheworldsplants.com.

9. M. Gagliano, M. Renton, M. Depczynski and S. Mancuso, "Experience teaches plants to learn faster and forget slower in environments where it matters," *Oecologia* 175 (2014): 63–72.

10. J. Böhm, S. Scherzer, E. Krol, S. Shabala, E. Neher and R. Hedrich, "The Venus flytrap *Dionaea muscipula* counts prey-induced action potentials to induce sodium uptake," *Current Biology* 26 (2016): 286–95.

11. S. W. Simard, D. M. Durall and M. D. Jones, "Carbon allocation and carbon transfer between *Betula papyrifera* and *Pseudotsuga menziesii* seedlings using a ^{13}C pulse-labeling method," *Plant and Soil* 191 (1997): 41–55.

12. S. Lewis et al., "Increasing carbon storage in intact African tropical forests," *Nature* 457 (2009): 1003–6.

Glossary

❊❊ ❊❊

agar: a substance extracted from seaweed, which solidifies into a media for growing seeds or tissue clusters in micropropagation.

annuals: plants that complete their life cycle in a year, from germination to setting seed, then dying.

anther: the pollen-bearing (male) part of a flower, often borne on a stalk (filament) at the top of the stamen.

binomial system: the universal system of naming species. Developed by the Swedish scientist Carl Linnaeus (1707–1778), it involves giving each species a two-part Latin name. The first is the genus to which it belongs and the second is its species name, e.g., *Ramosmania rodriguesi*.

bisexual flower: a flower containing both male and female organs in a single flower.

carpel: the ovule-bearing (female) reproductive organ in a flower. In many species, several carpels are fused together to form the ovary, style, and stigma.

carpellary appendage: an appendage growing from a carpel. The appearance, color, and shape of carpellary appendages are often used in waterlily classification for species identification.

chromosome: a threadlike structure made up of DNA, found in the nucleus of dividing cells.

clones: a group of genetically identical individuals reproduced from a single parent by nonsexual reproduction, *e.g.*, by taking a cutting.

colchicine: a chemical extracted from the autumn crocus (*Colchicum autumnale*), which is used to disrupt cell division and often causes chromosome duplication. The resulting offspring cease to be identical. In some cases it can render fertile plants sterile and sterile plants fertile.

critically endangered: a classification of vulnerability designated by the International Union for Conservation of Nature (IUCN), indicating flora and fauna that are "considered to be facing a very high risk of extinction in the wild."

cross-pollination: the transfer of pollen from one flower to another on a different plant.

cutting: a piece of leaf, stem, root, or bud, detached from a parent plant, that is induced to produce roots and later an independent plant. This is a clone, and genetically identical to the original parent plant from which it was taken.

dioecious species: plant species that produce separate male and female individual plants. Examples include ginkgo, juniper, and kiwifruit.

DNA (deoxyribonucleic acid): the hereditary material of most living things, containing the template for physical characteristics, growth, development, and functioning.

ecology: the branch of biology that studies the relationships between organisms and their environment.

embryo: a rudimentary plant within the seed of flowering plants that develops into the stems, leaf, and root.

endangered species: at high risk of extinction in the wild or in cultivation.

endemic: naturally occurring in one specific, usually restricted locality and nowhere else in the world.

endosperm: the seed's food supply until roots and shoots are formed.

epiphyte: a plant which grows on trees or other plants but does not take nutrients from them, so is not a parasite.

family: a unit of classification; families are divided into genera (see *genus*).

filament: part of a stamen; the filament supports the anther, which is where pollen develops.

form: a unit of classification that ranks below subspecies and variety.

gene: a unit of heredity that is transferred from a parent to offspring and is held to determine characteristics of the offspring.

genus: a unit of classification into which a family is divided in the binomial system. Genera are divided into species.

germination: a complex sequence of events by which a plant emerges from a seed.

habitat: an area in which a species lives, comprising the climate, soil type, topography, and other organisms.

herbaceous plant: a non-woody plant that usually dies down in winter.

herbarium: a collection of preserved, mostly pressed plants, cataloged and arranged systematically for the study of classification.

heterophyllous: having leaves of different shapes on the same plant.

hybrid: the offspring of sexual reproduction by two genetically dissimilar parents. These offspring are often larger, more vigorous, and healthier.

indigenous: a plant that originates in, or occurs naturally in, a particular area; native.

Latin name: the Latin description of an organism, usually made up of two words, *e.g., Ramosmania rodriguesi,* specifying the genus and species; see *binomial system.*

loam: a medium-textured soil with high proportions of sand, silt, and clay. Usually regarded as a good soil for cultivation.

Mascarenes: the group of islands in the Indian Ocean that comprises Mauritius, Rodrigues, and Réunion.

micropropagation: a method of plant propagation using extremely small pieces of plant tissue or a seed embryo.

mist unit: a piece of equipment, usually in the form of spray nozzles under plastic sheeting, that emits a mist of water droplets in order to propagate seeds or cuttings.

molecular: relating to or consisting of molecules (a group of two or more atoms). In botany, "molecular studies" often refers to the study of genes and DNA.

myxomatosis: a usually fatal disease in rabbits caused by the myxoma virus.

native: a plant of indigenous origin or growth.

neotropic/neotropical: one of the six biogeographical areas of the world, this comprises South, Central, and parts of North America, excluding the southernmost parts of Chile and Argentina.

nick: to remove a tiny section of seed coat with a knife, allowing the absorption of water and encouraging germination.

osmosis: the passage of molecules from a less concentrated solution to a more concentrated solution through a semipermeable membrane.

ovules: equivalent to human eggs; fertilized plant ovules lead to the development of seeds.

oxbow lake: a horseshoe-shaped body of water that is formed when the wide meander of a river is cut off, creating a freestanding lake.

Padrón pepper: from the Municipality of Padrón, in Galicia, northwest Spain (near Asturias). Often referred to as the Russian-roulette pepper, because although most of these peppers are mild, occasionally there is one that is ferociously hot.

pathology: the study of diseased, dying, or dead tissue.

petiole: a leaf stalk, connecting the leaf blade with the stem.

phenology: the study of recurring natural phenomena, especially in relation to weather, such as the annual appearance of the first butterfly or the blossoming of trees.

physiology: the scientific study of normal functions in living systems.

pollen: a coarse, often powdery substance made up of grains, which produce the male sperm cells (gametes) in seed plants.

pollen tube: a hollow tube that develops from a pollen grain when deposited on the stigma of a flower. It penetrates the style and allows the male sperm cells to reach the ovule.

pollination: the transfer of pollen from an anther to a stigma, in order to produce seeds.

pollinia: the lumped mass of pollen grains that is found in some plants, particularly orchids.

recalcitrant: used to describe seeds that can't be dried.

rhizomes: underground stems, often horizontal, lasting more than one growing season.

rooting hormones: chemicals, in the form of liquid, powder, or gel, that are applied to the base of cuttings to stimulate root production.

rootstock: see *stock/rootstock.*

scion: a shoot or bud cut from one plant for the purpose of joining it to another (the stock, or rootstock) when grafting or budding.

sepals: the outermost leaf-like organs of a flower, often green.

shaman: a person said to influence good and evil spirits for the practice of divination or healing. They often use hallucinogenic plants as part of their ritual.

species: a unit of classification into which a genus is divided; species can be divided further into subspecies.

spikelet: part of a grass flower head.

stamens: the male sexual part of the flower, consisting of a filament (stem) with anthers (pollen sacks) at the apex.

staminodes: hard, fleshy, infertile, modified stamens, which are often rudimentary or reduced.

stigma: the pollen-receptive area of a female flower, on the end of the style.

stock/rootstock: a stem and roots, onto which the scion is grafted.

style: part of the female sexual organ of a flower between the ovary and the stigma.

subspecies: a unit of classification into which a species is divided, formed by natural selection and generally geographically isolated populations. If isolation and differentiation continues, a subspecies can become a species.

taxonomist: a person who studies the classification of life forms and puts them into their correct categories.

taxonomy: the branch of science concerned with the classification of life forms; see *binomial system, family, form, genus, species, subspecies, variety.*

triploid plant: has an extra set of chromosomes; where there would normally be two, there are three. Triploid plants are generally sterile but are often quite vigorous.

type specimen: the single specimen on which the description and name of a new species, and therefore all other specimens, are based.

variety: a category in taxonomy that ranks below species and subspecies, and above form.

viable: capable of growth or reproduction.

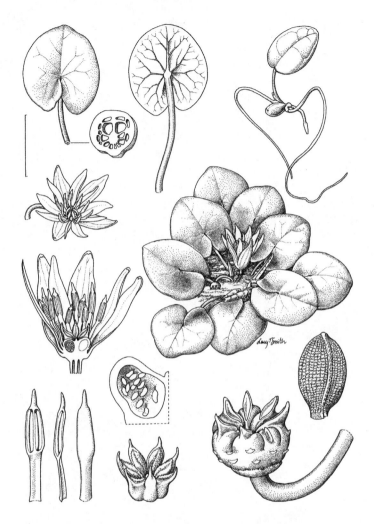

Botanical illustration of *Nymphaea thermarum*. FROM TOP TO BOTTOM,
LEFT TO RIGHT: *N. thermarum* pad from above; cross section of the petiole;
pad from below; seedling (with seed attached); flower from above; whole
plant; flower cross section; stamens (front, side, and back); carpel showing
ovule placentation; stigmatic disk and carpels; developing fruit with peduncle;
magnified seed.

(Drawing courtesy of Lucy Smith, first published in *Curtis's Botanical Magazine*, 27)